Technology and
Social Complexity

Technology and Social Complexity

Maurice N. Richter, Jr.
State University of New York at Albany

State University of New York Press
Albany

Published by
State University of New York Press, Albany

© 1982 State University of New York

For information, address State University of New York
Press, State University Plaza, Albany, N.Y., 12246

Library of Congress Cataloging in Publication Data

Richter, Maurice N.
  Technology and complexity.

  1. Technology—Social aspects. I. Title.
T14.5.R53   303.4'83   82-5683
ISBN 0-87395-644-3   AACR2
ISBN 0-87395-645-1 (pbk.)

10 9 8 7 6 5 4 3 2

303795

to my brothers
Marcel K. Richter
Wayne H. Richter

Contents

Acknowldgements

This book was largely written during a sabbatical leave from the Sociology Department of the State University of New York at Albany. I am grateful to the administrators who approved the arrangements for this leave, and especially to Nan Lin, Chair of the Sociology Department.

During my sabbatical, from January through August 1981, I was a visitor in the Graduate Program in Science, Technology, and Public Policy at George Washington University. I am most grateful to the faculty and staff members in this program, and especially to its director, John M. Logsdon, for the hospitality that made my visit there a pleasant one and for access to resources and facilities that aided me considerably in this work.

The chapter on Chinese technology is based partly on observations that I made during a trip to China in February 1979, while I was holding a National Endowment for the Humanities fellowship and was in residence at the American Enterprise Institute for Public Policy Research. For the financial support that made that trip possible, I am indebted to AEI and especially to William J. Baroody, Jr., President; Gary L. Jones, former Vice-president for Administration; and Robert J. Pranger, Director of Fellowship Programs.

For helpful comments on the manuscript, I am indebted to Judith R. Blau (SUNY–Albany). I have also received helpful advice on specific points from William N. Fenton, Morris Finder, Arnold W. Foster, Richard H. Hall, Nan Lin, and Walter P. Zenner (all of SUNY-Albany), and from Howard S. Klein (China Consulting Associates), Robert W. Lamson (National Science Foundation) Peter Nelligan (University of Hawaii) and Edward F. Wente (University of Chicago).

And once again, as in three previous works, it is a pleasure to acknowledge my intellectual debts to three of my former teachers, Gerard DeGre, Anselm L. Strauss, and Edward Shils, and the encouragement that I have continually received from my parents, Maurice N. Richter and Brina Kessel Richter.

No one mentioned above is responsible for any errors or inadequacies that may be found herein, or for any of the views that I have expressed, some of which may turn out to be highly controversial.

1
Introduction

About the year A.D. 40,000,000, an observer on Venus prepared a history of human life as it had existed on the recently destroyed planet Earth. By the year A.D. 8,000,000, he reported, many spacecraft had been sent from Earth to that planet's only moon. All but two of these had been destroyed en route by friction within Earth's atmosphere, by collision with meteorites, or by crash landing on the moon. Crews of the two spacecraft that landed safely had mapped the side of the moon that always remains hidden from Earth and, no return to Earth being possible, had died on the lunar surface. J. B. S. Haldane published this imaginative account in *Possible Worlds* in 1930.[1] Thirty-nine years later, rather than eight million years later, men had landed on the moon and returned safely home; no one had died en route; and the moon's hidden side had already been thoroughly mapped by cameras in lunar orbit.

As a child in the early 1940s, I was fascinated by Haldane's fictionalized predictions, and by Clifford C. Furnas's *The Next Hundred Years*. Appearing in 1936 this book did not mention space flight as a possibility by the year 2036 but did comment on many other matters, including the need for faster airplanes, suggesting that "the ultimate attainable speed . . . might well approach 500 miles per hour."[2]

Furnas was not the only person to underestimate the future of aviation. Wilbur Wright, more famous as coinventor of the airplane than as a predictor of its future development, reportedly thought in 1909 that planes would not be used for bombing in wartime because they would have to fly at one thousand feet or higher "to escape shellfire" and "at that height accuracy would be impossible."[3] And in 1903, only a few months before the great achievement of the Wright

1

brothers, one observer speculated in a document entitled "The Out-look For the Flying Machine," that we may "be ultimately forced to admit that aerial flight is one of that great class of problems with which man can never cope, and give up all attempts to grapple with it."[4]

Such underpredictions of technological achievement have been balanced by overpredictions. We are still waiting for the "machine three fingers high and wide" by which "a man could free himself and his friends from all danger of prison and rise and descend. . . ."[5] that Roger Bacon told us was possible seven centuries ago. There have also been no visible results yet from the claim made in the Soviet Union in 1932 that "we are solving the problem of heating Siberia."[6]

Predicting general social conditions under which technological (including scientific) innovation is most likely to occur is as problematic as predicting particular innovations. Thus, an American sociologist has argued that "liberal society is most nearly favorable . . . to modern science,"[7] whereas a Soviet author has claimed that "only under socialism and communism are all obstacles to scientific and technological progress eliminated."[8]

The record with respect to predictions of *effects* of new technology is similarly unimpressive. For example, one might have reasonably predicted that the substitution of jets for propeller planes and the greatly increased reliability of jets would make it possible not only for travelers to have a first lunch in London and a second but "earlier" lunch in New York on the same day, but also for travelers to find themselves moving at great speed toward remote and unexpected destinations as hijacking victims. There did not seem to be much awareness, however, as the world crossed the threshold of the jet age, that the comparative slowness and unreliability of aviation in the earlier propeller age had provided crucial and now-disappearing protection against a spectacular extension of terrestrial crime into the skies.

Difficulties encountered in predicting effects of new technologies are nicely illustrated by the history of predictions concerning effects of new military technology on the destructiveness of war. The German military theorist Karl von Clausewitz (1780–1831) saw improvements in military strategy as leading to more precise focusing of military action on an increasingly narrowly defined range of targets and hence to a style of warfare less destructive than the styles of the past: "Civilized people," in contrast to "savages . . . do not devastate towns and cities . . . because their intelligence exercises greater influence on their mode of carrying on war."[9] In 1621 John Donne claimed that new weapons were shortening wars and thus reducing casualties: "Artillery, by which warres come to quicker ends than heretofore," meant that "the great expence of bloud is avoyded; for the numbers of men slain now, since the invention of Artillery, are much lesse than before, when the sword was the executioner."[10] Still others have seen new weapons as so terrible in their potential effects that they would discourage the outbreak of war altogether. Thus, Benjamin Franklin expected that manned balloons might "give a new turn to human affairs" by "convincing sovereigns of the folly of war"[11], and James Harvey Robinson observed in 1902 that "repeating rifles and increasingly powerful explosives have made war so destructive that statesmen are more and more reluctant to suggest a resort to arms."[12]

The general relationship between technology and society is a topic on which we find an extreme diversity of perspectives. At one pole we find technological determinists such as Jacques Ellul[13] and Leslie A. White, for whom social systems are "subsidiary to technological systems."[14] The technological-determinist perspective was incorporated into the motto of the 1933 Century of Progress Exposition in Chicago: "Science Finds—Industry Applies—Man Conforms."[15] Decisive rejections of material-technological determinism come from political leaders who seek to bring about fundamental transformations of their societies, who have only comparatively backward systems of material technology at their disposal, and who must accordingly rely heavily on human muscles and especially human willpower. China's Mao Zedong (Mao Tse-tung in the old spelling), insisted for example, that *people* (i.e., the "masses") were the decisive element in human history and that the masses, properly inspired and organized, could prevail in opposition to their enemies even if they did not have, as their enemies had, access to the most highly developed material technologies. Between the technological determinists and their most extreme opponents, we find diverse intermediate positions.

The contrasting perspectives that exist in the study of the technology-society relationship are matched by considerable diversity in the characteristics of those who have concerned themselves with this topic. The latter include technologists and scientists, dealing with their own (or other) specialties; novelists (including science-fiction writers); philosophers; political ideologists and politicians of various sorts; government officials; representatives of diverse commercial interests; and social scientists representing several disciplines—economists, political scientists, historians, anthropologists, and sociologists.

Among the social sciences, one discipline, sociology (which happens to be my discipline), occupies a particularly crucial position in the study of technology in society. This, however, is merely a consequence of the crucial position of sociology among the social sciences generally. Economics and political science are concerned primarily with particular social institutions (i.e., with economic and political institutions respectively) rather than with society as a whole. Those portions of anthropology that are concerned with society are similarly focused on one particular aspect of it, the aspect that we call "culture." (Other portions of anthropology deal with biological characteristics of the human species rather than with human society as such.) History, like sociology, covers all parts of human society, but history focuses on detailed descriptions of unique past events (and, to a limited extent, on recurring patterns of events). Sociology appears to have a dual role within the broader field of social science. First, it occupies whatever territory happens to be left over after all other social-science disciplines have staked out their claims. This means, for example, that sociology encompasses the social-scientific study of such "leftover" topics as religious and family institutions, but *not* economic or political institutions, which are claimed by economics and by political science respectively. *This* role of sociology tends to give it a scattered, disorganized character. But sociology also has a second, quite different role: it encompasses the study of relationships between and among various aspects of society (e.g., between economic and political institutions) and the study of the organization of society as a whole. Although economics, political science, history, and anthropology each makes a specialized contribution to the study of the relation between technology and society, only sociology provides a potential basis for integrating these contributions and for analyzing the technology-society relationship in general.

There are many divergent styles of sociological research and theorizing. I shall attempt to briefly describe those aspects of my own style that give this study its distinctive character.

1. I am as concerned as anyone else about the unprecedented opportunities and dangers that contemporary technology entails, and about moral and ethical issues that technological progress has rendered particularly acute. However, I do *not* focus here directly and primarily on such contemporary concerns. Rather, I seek to view technology in what might appear to be a comparatively detached way. Underlying this approach is a basic assumption: some things can be done most effectively if one does not try too hard to do them. Going to sleep is one of these: if you had an urgent reason to want to fall asleep immediately, without drugs or other artificial means (for example, if someone offered you a huge sum of money if you did so), your heroic efforts would

probably keep you awake and thus be self-defeating. Sometimes, efforts to understand an intractable problem can be self-defeating in much the same way: in concentrating on the problem, we may shut out of our mind all "irrelevancies"—and in doing so we may accidentally shut out the key to the problem's solution. If the problem is really a stubborn one, we might do better by adopting a more relaxed attitude, playing with our minds around the edges of it, and trying to view it in as broad a context as possible. A general understanding of the relation between technology and society, with much-talked-about practical and moral issues de-emphasized, may thus contribute usefully even though indirectly—or perhaps we should say "usefully *because* indirectly"— to a clearer understanding of these issues.

2. I assume that retrospective views of the historical record tend to be systematically selective in various ways and that this selectivity has a generally distorting effect on interpretations of historical trends. I am thus highly skeptical even about the existence of trends that are generally taken for granted as too obvious to require proof. In the 1970s, while inflation appeared to be reaching alarming proportions, I startled my students by insisting that the existence of inflation is an artifact of our methods of measurement. If one compares the cost of a hospital stay today with its cost early in this century, when "about one dollar a day" was described as "the expense of treatment, board, etc." at the Cook County Hospital in Chicago,[16] then of course medical costs will seem to have zoomed upward. If, however, we take certain ailments (tuberculosis, typhoid fever, pneumonia, polio, etc.) and compare the cost of preventing or curing each of these today with the cost of earlier attempts at prevention or cure, *then* the price of caring for our health will appear to have descended sharply. I have startled my students also by insisting that, if we confine our attention to wars already fought rather than to hypothetical future wars, the currently popular idea that warfare has become increasingly destructive is questionable: it is not at all clear that the world wars of our century have been more devastating to Western society than the Peloponnesian War was to the society of ancient Greece twenty-four centuries ago. My skepticism about what are commonly regarded as "obvious" trends pervades the present study, leading me sometimes to "move backward" by raising questions about matters that are commonly recognized as already settled, rather than "moving ahead" toward new questions and new answers to questions currently under public discussion.

3. I am concerned not only about systematic distortions of historical trends in careless retrospective interpretations but also about

distorted interpretations of social phenomena resulting from the failure to take macroscopic perspectives into account. In my previous studies of science, I have been impressed by a striking difference between the way in which science appears to be heavily "controlled" by social pressures when one examines numerous particular contexts in which scientific research is undertaken and the way in which science at a global level appears comparatively free from social control. This paradoxical contrast emerges from the fact that social controls over science tend to be localized and locally variable: what cannot be done in one time and place can very likely be done in another. Science has progressed largely by flowing around localized obstacles (although sometimes at heavy cost): Galileo published in Holland when he could not do so in Italy; physiology flourished in France when antivivisection movements and other problems made its flourishing difficult in early nineteenth-century England; scientists fleeing from Hitler stimulated science outside of Europe; and genetics, when suppressed in Stalin's Russia, thrived elsewhere. What is true of science appears true of technology also: localized observations of the technology-society relationship are likely, by themselves, to give us a distorted view of this relationship as a whole. For example, from a comparatively *micro*scopic point of view, social pressures often appear to sharply control the appearance and utilization of technological innovations, whereas from a more *macro*scopic perspective, technology appears to have had a relatively (although of course nowhere near "completely") autonomous growth.

These assumptions and concerns suggest a comparatively broad definition of "technology." As defined here, the term encompasses not only such recently emerging wonders as trips to the moon, hydrogen bombs, genetic engineering, and computers which some fear will eventually enslave us, but also toothbrushes, doorknobs, and wheelbarrows, as well as simplified spelling systems, arrangements for bureaucratic coordination, military draft lotteries, advertising strategies, and even sticks that chimpanzees use to catch termites. I seek to show that such a wide range of phenomena can be analyzed coherently within the framework of the same concept and to provide a broad historical and comparative perspective within which our exciting, promising, and deadly new technologies can be more clearly understood.

2
Technology and Its Forms

Technology Defined

According to one view, technology is not merely a human phenomenon but rather is found also among diverse lower-animal species; for example, the term has been applied to the nest building of birds and the dam building of beavers.[1] At the opposite extreme, technology has been defined as "the systematic application of scientific or other organized knowledge to practical tasks,"[2] and this definition presumably confines it to those human societies that have attained comparatively high levels of sophistication. Here, technology will be conceived in a way that is intermediate between these two extremes, as a phenomenon universal among human societies (rather than confined to only some of these) but not encompassing the characteristic activities of birds or beavers.

Elsewhere I have defined technology as "knowledge of any technique for achieving specific objectives"[3] and have distinguished science from technology in this sense as follows:

. . . "The scientific method" of investigation is itself an element of technological knowledge in our society: it is knowledge about a way of achieving the objective of scientific progress. Knowledge of natural laws, acquired through use of the scientific method, may or may not be technological, depending on the situation: it is technological if recognized as a means for achieving some further objective such as control over nature, but not if it is valued merely "for its own sake." Science itself is not part of technology as defined here: technology is a kind of knowledge, while science is a

7

search for knowledge (a search which *utilizes* previously acquired technology, including the "scientific method" as a technological element, and which *produces* knowledge which is *sometimes* technological.[4])

These words were written in the context of a study that focused on science and that discussed technology only incidentally for the purpose of distinguishing it from science. In *that* context a definition of technology as a kind of knowledge was appropriate, because the knowledge aspect of technology bears a closer relationship with science than do its various other aspects. However, when our attention is focused primarily on technology rather than on science, a more inclusive definition seems in order. A country might have citizens who "know" how to put a satellite into earth orbit but might nevertheless not have the facilities or resources to accomplish this. If we define technology merely as a kind of knowledge, then we would have to say that that country possesses the technology of space flight even though it cannot actually put an object into space. We can avoid such awkwardness by recognizing that technology involves more than mere knowledge. In fact, for present purposes it will be most convenient to abandon the conception of technology-as-knowledge, and to define technology to encompass *tools and practices deliberately employed as natural (rather than supernatural) means for attaining clearly identifiable ends.*

When technology is defined (as I previously defined it) as a kind of knowledge, science cannot be classified as a kind of technology, as long as science is understood as a *search for* knowledge rather than as knowledge itself. By contrast, our present definition of technology as encompassing certain "tools and practices" enables us to recognize science-as-a-search-for-knowledge as a kind of "practice" included within the "technology" category, except when science is undertaken only "for its own sake."

This definition excludes anything that is "an end in itself" rather than "a means to an end." For example, anything that is artistic *rather than* functional (as distinct from being *both* artistic *and* functional) would be essentially nontechnological, although technologies of various sorts may be used in producing these. (An automobile is an item of technology, and a work of art is not, although technologies are employed in producing both the automobile and the work of art.)

In excluding from "technology" anything that is not a means to an end, we make *no* exception for phenomena that perform useful or even essential functions if these functions remain unknown to relevant participants and thus do not constitute "means" in relation to "ends."

Sex among lower animals provides a good example: it serves an essential reproductive function; but this function is presumably unknown to the participants, who engage in sex merely for the immediate pleasure involved. Similarly, when a bird builds a nest, it is presumably motivated by satisfactions inherent in the nest-building process itself rather than by anticipation of the joys of laying eggs and rearing young that are to follow. A genuine means–ends relationship, and hence genuine technology, may be more reasonably recognized in cases involving more intelligent animals and briefer sequences of events, for example, the use by chimpanzees of tools that they specially prepare to catch termites. One observer saw chimpanzees in the wild search for twigs or blades of grass of the appropriate size and stiffness, pulling strips from blades of grass if they were too thick, and stripping side branches and leaves from the twigs. The chimpanzees would then search for termite nests and carefully insert their implements. The termites inside would bite, and remain attached to, an inserted twig or grass blade, which the chimpanzee would then remove slowly and gently to avoid knocking them off.[5]

In human society we find some activities and patterns that are similar to sex among lower animals in that they are maintained because they are immediately satisfying in some sense and that also serve functions that are latent in the sense that participants are not generally aware of them. Some aspects of the American family system illustrate this kind of nontechnological phenomenon. This system (unlike its counterparts in some other societies) is organized around the "nuclear family," consisting of husband and wife and their children; larger kinship units such as lineages and clans are relatively unimportant in the United States. This arrangement creates some problems when a marriage is broken by death or divorce, and when young children have no larger family unit which they can rely on to take over parental responsibilities after the nuclear family thus falls apart. However, these problems are compensated for by a huge advantage: a nuclear family can move relatively easily from one region of the country to another in response to geographically shifting employment opportunities, whereas larger family units would be less mobile and thus would tend to reduce the geographical mobility of their individual members. The American family system contributes in this way to economic arrangements that require a mobile labor force. However, we should not call this family system a technological "means" for facilitating labor mobility. No one "planned" our family system with this purpose in mind. Rather, the system evolved spontaneously through trial and error over a long period of time. Furthermore, support for this system is directly motivated by the belief that it is fundamentally right and proper and by immediate satisfaction with it rather than by utilitarian means–ends calculations.

There is still another facet to the boundary of the technology concept and hence another type of nontechnology. People may employ various practices in pursuit of clearly identifiable ends, but through supernatural means, for example, through prayer calling for divine intervention. Our definition of technology requires that natural means be employed: we shall not allow, as at least one other investigator has allowed, for a "technology of prayer."[6] However, our definition does not specify that the natural means that are employed must be effective. An effectiveness requirement is incorporated into some familiar definitions: for example, "the totality of methods rationally arrived at and having an absolute efficiency . . . in every field of activity,"[7] and "any set of standardized repeatable operations that regularly yield predetermined results."[8] Such a definitional requirement is sometimes difficult to apply. For example, some medical treatments are effective only because people expect them to be, and some (including bloodletting as a generalized treatment for many ailments) have been found to be *in*effective only after centuries of use.

One more example will clarify further the boundaries of the technology concept as here defined. In a primitive tribe, rain dances may be held for the purpose of producing rain. An outside sociological observer from a more advanced society might conclude (1) that the rain dance does not really produce rain, (2) that this dance nevertheless performs a useful function by sustaining morale and/or reinforcing tribal unity, (3) that these actual functions that the rain dance performs are "latent" (i.e., not clearly recognized by the participants, who believe that the only function of the dance is to produce rain), and (4) that the rain dance has persisted as an element of tribal tradition only because it has been performing its latent functions.[9] Because the functions that the rain dance actually performs are unrecognized by the participants, they are not "ends" in relation to which the rain dance could be properly classified as a "means." The rain dance *is* a means in relation to the "end" of producing rain. If, as we would assume, no rain is actually produced and the means is thus ineffective, this does not rule out classification of the rain dance as a "technological" item, because actual effectiveness is not required by the definition of technology. However, application of the technology concept *is* ruled out by another consideration: by the fact that the "means" in this case is supernatural, entailing an appeal to spiritual entities.

In real life, the facts are often not as clear-cut as we imagine them to be in this hypothetical example. We must thus allow for unclear borderline situations, for phenomena that may be "semitechnological" in any of several ways.

Varieties of Technological Phenomena

Technology includes material tools available to a society, for example, in our society, airplanes, computers, and nuclear weapons, as well as wheels, boxes, toothbrushes, and pianos.[10]

However, human organizations may be used as "tools" just as material objects are. An army may be a tool in the hands of a general or a ruler, just as a sword or a gun is a tool in the hands of a soldier. Furthermore, various similarities between material and organizational tools justify grouping them together. Both are subject to similar processes of invention: the corporation was "invented" just as the automobile was. There are sometimes remarkable similarities in their internal functioning: the standardization of parts in modern machinery, which makes possible the replacement of components with "spare parts" that ideally function essentially as the original parts did, is paralleled by the standardization of roles in bureaucratic organizations, which often renders individual personnel replaceable in much the same way. Underlying similarities between material and organizational tools are reflected in our language by terms that cut across the distinction between them: we have *political* "machines" as well as material ones. In fact, Lewis Mumford has suggested that machines in which the component parts consisted of *people* were the *first* machines to emerge in history and that they served as prototypes for the construction of machines composed of inanimate materials.[11]

A third type of "tool" may also be recognized. A new system of writing or calculation, a new way of classifying phenomena, may sometimes produce dramatic new opportunities. The transition from Roman to Arabic numerals provides a clear example: one need only think of how much easier it is to multiply 12×17 than to multiply XII $\times$ XVII. We thus have "symbolic" as well as material and organizational tools.

Technology has been defined here to include both certain tools and certain practices. Some practices are closely linked with tools (e.g., surgical practices with surgical tools), whereas others are not. A type of technology that does not necessarily involve tools (although certain specific forms of it do) is the type that is concerned with *persuasion* in a broad sense: encompassing the informal teaching of children by their parents and peers, formal education as conducted in schools, psychotherapy, hypnosis, and the manipulative persuasive tactics of commercial advertisers and of political and religious propagandists.

It is possible to have not only technologies that do not entail tool using, but also technologies that consist essentially of refraining from doing certain things. If modern medical research should discover that a given disease can best be cured by having its victims refrain from eating certain foods, then this cure—which fits within our definition of technology as long as "refraining from" doing something can be classified as a "practice"—involves no tools and no positive activity of any kind. In the eighteenth century, German peasants made what to them was an important discovery in agricultural technology, which similarly involved merely refraining from doing what they had been doing previously and entailed no positive action. They discovered that potato crops did not need to be harvested before winter and stored (a procedure that made the stored crops easy for marauding armies to requisition). Potatoes, they found, could be left lying in the fields during the winter without spoiling and collected from the fields only as needed; left in the fields, the potatoes could not be so easily requisitioned or seized.[12]

The recognition that material technology is not the only type and the allowance for a variety of nonmaterial technological forms has certain implications that may be noted as follows:

A distinction has been made between technologies and "supporting systems" which are defined as "legal and economic arrangements through which . . . technologies become available and are subjected to social control." Thus, "the automobile and the highway network comprise a technology," whereas "rules of accident law, automobile insurance schemes, and traffic policemen are components of the corresponding supporting system."[13] The approach to be followed in this book involves two modifications in this distinction. First, according to our conception of technology, this distinction can be made only on a contextual basis: what is part of a supporting system in one context may be a central technological element when the context shifts. Thus, automobile insurance schemes may be part of the "supporting system" for the automobile as an element of transportation technology; but when protection of the financial interests of people involved in auto accidents becomes a social end, auto insurance arrangements may constitute a technology employed in pursuit of that end. Second, a "supporting system" as conceived here need not be merely a set of "legal and economic arrangements": one kind of material technology may be part of the supporting system for another. Thus, the supporting system for the automobile in the United States should be recognized as including not only insurance but also large tankers that transport oil across the oceans, and tools used in auto-repair shops.

The present definition of technology is broad enough to encompass even some developments that are commonly considered to be "antitechnological." The Chinese Communists during the time of Chairman Mao's leadership made a virtue of necessity by stressing the importance of people and the relative unimportance of advanced material technology in achieving various societal goals. They carried this approach even to the point of organizing people to catch millions of birds with their bare hands and with such simple equipment as hand-held nets. One morning in 1957, the population of Beijing was organized in a "war against the sparrow," which commenced at 4:45 A.M. The plan was to station people everywhere—on roofs, in the branches of trees, in streets, in fields—so the sparrows when flushed from their resting places all over the city would find no place to land and would eventually drop exhausted to the ground, where they could be picked up by hand and piled in trucks that circulated through the streets.[14] This method contrasts strikingly with the way in which a few aircraft have been used to spray deadly chemicals on millions of roosting blackbirds in the United States. It involved a rejection of advanced *material* technology in one respect, but not a rejection of technology itself as here defined; in fact, it involved an interesting form of *organizational* technology (i.e., a way of coordinating behaviors of large numbers of people in the performance of social tasks).

Technology and Tradition

The present conception of technology may be further clarified by examining two manifestations of it, neither of which is ordinarily thought of as having anything to do with technology at all.

The United States Constitution

More than two decades before the Declaration of Independence, Benjamin Franklin, who knew about the Confederation of the Iroquois Indians in some detail, pointed to that Confederation as an example that the British colonists in North America might well emulate:

> It would be a strange thing if Six Nations of ignorant savages should be capable of forming a scheme for such an union, and be able to execute it in such a manner as that it has subsisted ages and

appears indissoluble, and yet that a like union should be impracticable for ten or a dozen English colonies, to whom it is more necessary and must be more advantageous, and who cannot be supposed to want an equal understanding of their interests.[15]

The United States Constitution of 1789 was an attempt to do what the Iroquois had already apparently done, and what the city-states of ancient Greece, despite their admirable achievements in other respects, had failed to do: to create a stable, voluntary union of separate, culturally similar (but not culturally identical) states in which each would gain from the strengths of the others and in which each would enter on an essentially equal footing. It was an attempt not merely to provide a basic document for the government of a nation, but to create that nation while doing so, through a recombination of units that had earlier been linked together more loosely in another form, as British colonies, and that had broken away from the British Empire together.

But although the Constitution thus united the thirteen original American states (and paved the way for other states to join them), it did so in a way that decisively rejected the concept of a highly centralized polity. Instead, it reserved certain functions for state governments and/or for "the people," and instituted at the national (Federal) level a separation among executive, legislative, and judicial branches, with each of these made strong enough to impose major restrictions on the power of each of the others and with all of them sharing some of the same functions. For example, the legislative function is performed not only by the legislative branch of government (Congress) but also by the executive branch, with the president proposing and sometimes vetoing legislative acts, and by the judicial branch, which can reject legislative acts on the ground that they are "unconstitutional".

The historical significance of this can be understood only by contrast with corresponding European developments. Prior to the rise of major national states in seventeenth-century Europe, there had been considerable political fragmentation, not because this condition was deliberately planned by anyone, but rather as an unintended, unplanned consequence of stalemates among competing centers of power. In the seventeenth century, although Europe as a whole remained politically fragmented (as it has generally been, and as it remains today), nevertheless political power became strongly centralized within particular European states that were unified under authoritarian rulers. This development was symbolized most clearly in France under Louis XIV (1638–1715), who reportedly proclaimed, "L'état c'est moi!" [I am the

State!]. In England a system was evolving in which power also came to be highly centralized, although in a legislative body (parliament) rather than a monarch. The United States Constitution rejected the European movement toward national political centralization and aimed instead to preserve, in officially established form, a decentralized governmental structure somewhat reminiscent of those found in Europe (especially England) in earlier days.

Highly centralized systems emerged in Europe largely through deliberate construction, and hence "technologically." The "separation of powers" in Europe prior to the rise of strong national states was not a technological innovation but primarily a historical accident; the separation of powers in the United States was a deliberate product of a written constitution and a clear example of organizational technology.

The separation-of-powers arrangements incorporated into the United States Constitution have been widely imitated in other institutions of American society. Although the Constitution imposes only very loose restrictions on the forms that *state* governments may assume (requiring only that each state have a "republican" form of government), nevertheless a remarkable uniformity among state governments has emerged, in spontaneous, voluntary imitation of the United States Constitution. No state has instituted a highly centralized European-type system, although the Constitution would not appear to prohibit this.[16] Similarly, a separation-of-powers arrangement reflecting Constitutional principles has been instituted in many American universities, which have legislative bodies ("senates") in which administrators are not necessarily members, in contrast to the system prevailing in an English university that I visited, in which all department heads had university senate membership automatically (just as the British prime minister and other ministers are members of the national legislative body [parliament]).

To the extent that the United States Constitution has come to be taken for granted and valued "for its own sake," with its principles upheld (in the constitutional context itself or in the contexts of other institutions) through blind adherence to tradition rather than as an outcome of means–ends calculations, to that extent the Constitution has been losing its original technological character. We have here a reminder of a possibility not often considered: we all know that an established technology may be abandoned and may disappear, but it is also possible for an established technology to continue to exist while ceasing to be fully "technological."[17]

The Achievement of Sequoyah

The historical development of writing has involved a transition from picture symbols to syllabaries (in which each symbol ideally stands for a syllable) and from syllabaries to alphabets (in which each symbol ideally stands for a phoneme, which is a unit of sound smaller than a syllable). But systems of writing commonly diverge from "ideal" patterns, and sometimes for good reasons.

Given the way English is actually pronounced (a condition that we cannot reasonably hope to change to any considerable extent through deliberate planning), we would not want our spelling to match pronunciation patterns precisely. Identical spellings for quite different words that happen to be pronounced the same way (*cite, site,* and *sight,* for example) could make comprehension in reading much more difficult, even if it made the task of spelling easier. And if we wrote *cats* and *dogz,* which perfect correspondence between pronunciation and spelling would require, this could obscure for the reader the fact that plurals are similarly intended in both cases. However, discrepancies between spelling and pronunciation are sometimes irrational obstacles to learning: we would be better off with such spellings as *enuf, thru,* and *tung.*

The Norman conquest of England in 1066 was followed by a period in which the English language was subordinated to Norman French, and English writing was temporarily eclipsed. When writing in the much-changed English language reappeared strongly three centuries later, spelling was apparently more consistent with pronunciation than it is today. The growing inconsistency between spelling and pronunciation has been caused primarily by shifts in pronunciation while spelling has remained comparatively stable and by imports of foreign words that retained their original spellings while receiving anglicized pronunciations.

In the early nineteenth century, the Cherokee Indian Sequoyah sought to drive the white men out of North America and, as a step in this direction, sought to make available to his people the white man's secret of writing. "Much that red men know, they forget," Sequoyah reportedly said; "they have no way to preserve it. White men make what they know fast on paper like catching a wild animal and taming it."[18]

Sequoyah did not know English or any language other than his native Cherokee. He was illiterate except in the script that he himself invented. After tedious experimentation with picture symbols and with writing systems in which each symbol represented a *word,* he finally rejected these as too complicated to provide a basis for mass literacy for his people. He came to believe (erroneously) that in the white man's writing system each symbol represented not a word but a syllable. He thus constructed what *we* would call a syllabary, although Sequoyah himself had no abstract knowledge of the differences among various types of writing systems. After several years of determined and lonely effort and after enduring ridicule from fellow Cherokees who thought he was crazy, Sequoyah was finally ready, in 1821, to demonstrate his system of "talking leaves."

Sequoyah's syllabary was specifically tailored to the features of the Cherokee language and thus did not have general applicability beyond the boundaries of his own tribe. Within the Cherokee context, it permitted exceedingly rapid acquisition of the ability to read simple messages; bright, energetic, and strongly motivated Cherokees apparently became "literate" within a few weeks and sometimes even in a few days.

A deliberately constructed writing system is maximally "technological" at the time of its construction. If it remains frozen while relevant circumstances (e.g., pronunciations) change, it will still presumably involve a technological aspect insofar as it is a "means" for achieving certain communicative "ends," but its technological status will be diminished by its growing entanglement in unquestioned tradition.

English writing was relatively strongly technological when its spelling was closely linked to pronunciation. It has been gradually drifting away from this condition. Sequoyah's achievement represents an episode of *purely* technological innovation emerging suddenly and sharply against a long historical background of what had become "semitechnology."[19]

In discussing the definition of technology in this chapter, I have unavoidably focused largely on phenomena that lie along the various boundaries of that concept, phenomena that are semitechnological in various ways. In discussing the relation between technology and society in succeeding chapters, I shall quite reasonably focus more on phenomena that fall clearly within the boundaries of "technology." In another respect, however, the style of this chapter will be continued in the remainder of the book. Many studies of "technology and society" emphasize material technology to the neglect of technology in its organizational and symbolic forms, or focus on dramatic technologies (space flight, computers, nuclear energy) to the neglect of more ordinary forms (toothbrushes, doorknobs), or concentrate on those technologies that raise issues of policy. I have sought to avoid such imbalances in this chapter and shall continue to try to avoid them, in an effort to clarify the way in which human society is related to technology as a whole.

3
Society and Its Evolution

The Concept of Society

In a work devoted to integrating the study of technology with the study of society, a chapter introducing the concept of technology may appropriately be followed by one in which the concept of society is discussed.

Humans are social animals, living in cooperatively organized groups. Cooperation *within* groups facilitates, and is facilitated by, competition and conflict *among* groups, but there is considerable intragroup competition and conflict as well.

Human groups are organized at diverse levels, with larger groups containing smaller ones and with some groups cutting across the boundaries of others (e.g., family and professional groups with memberships geographically dispersed and thus divided among different communities or nations). Societies are the most inclusive groups that maintain strong cohesion. California is not a society but merely a part of one; the human world as a whole lacks sufficient cohesion to constitute a single society and is best regarded as containing numerous societies within itself; the United States may reasonably be called a society. In the contemporary world, politically independent nations are the most "societylike" of all social units, but two precautions must be taken in identifying "society" and "nation" as equivalent terms applicable to any country.

1. Some countries display characteristics of nationhood and of society more than do others: Canada, which is divided between English-speaking and French-speaking populations and highly dependent

economically and militarily on the United States while also until 1982 politically subordinate to Britain in certain respects (prior to 1982 the Canadian Constitution could be amended only with approval of the British Parliament), is less nationlike and less societylike than the United States. Some countries that are very much more fragmented and dependent (e.g., Lebanon as of the summer of 1982) are hardly recognizable as discrete nations or societies.[1]

2. Nations as we know them are characteristic only of our own historical era; human society has appeared in nonnational forms in the past and may do so again in the future. In fact, nations today appear to be losing their societal statuses, with some societal functions devolving on lesser groups within them, and others being taken over by newly emerging structures at a global level—but these trends need not concern us at this point.

Human society has diverse aspects and may be analyzed in a variety of quite different ways. Societies may be described in terms of the knowledge and technologies available to them. A societal population may be described in terms of size and growth; birth and death rates; life-cycle stages through which its members pass; the geographical concentration and dispersion of its members and their intra- and intersocietal migrations; similarities and differences among its members. Some similarities and differences are biologically determined (sex, age, race); others are determined or influenced by social tradition or *culture*. Society needs certain similarities among its members (in order that they can communicate with each other and cooperate on the basis of shared interests or values) *and* certain differences among them (in order that a division of labor can exist and excessive competition for the same resources can be averted). Along with the patterning of similarities and differences among people, we find a patterning of relationships among them, involving such forms as cooperation, competition, and conflict. Furthermore, any human society will incorporate a number of systems or institutions, each focusing on a different general societal problem: economic institutions are concerned with production and distribution of goods and services; political institutions with societal decision-making and enforcement of societal decisions; religious institutions with maintenance of morale and morality; educational institutions with preparation of people for adult roles and specialized occupations. In comparatively primitive societies, these diverse systems are commonly centered to a large extent around the family, whereas in modern societies institutions are more highly differentiated and specialized, and the family, although remaining important, has a narrower range of functions.

Changes in society should be distinguished from processes that constitute part of society's routine functioning: when individual people pass through stages in the life cycle from birth to death, their own individual situations are changing; but this does not entail any change in society as an overall system. Social changes may be "random" (as when a language becomes differentiated into dialects), cyclical (as in the well-known "business cycle" and cycles in fashion), or unidirectional (as illustrated by the progress of science).

Unidirectional change does not rule out a major collapse that eliminates the effects of a long period of progress; it merely rules out a restoration of earlier conditions through steps that reverse those previously taken. It means that after climbing up the ladder of progress, we cannot *climb* down again although we may *fall* down or the ladder may break and drop us. Unidirectional scientific progress thus does not rule out a catastrophic destruction of civilization with loss of accumulated scientific knowledge: it merely means that we cannot make an orderly retreat from our present level of scientific knowledge back through earlier levels in a reverse sequence. Even with this qualification, however, the distinction between cyclical and unidirectional change is sometimes hard to apply. An economy (or another aspect of society) may manifest long-range unidirectional growth while also passing repeatedly through cyclical phases, and changes that are basically unidirectional may involve reversions to earlier conditions in certain limited respects (as when modern industrial societies restore egalitarian traditions of primitive times, traditions that had been abandoned as societies intermediate between primitive and modern acquired exalted rulers and degraded masses).

Unidirectional social change is usually referred to as "development" when it follows a predictable sequence of stages comparable to those through which a child passes in growing up or when it is deliberately planned with some form of modernization as the intended end. The term "evolution" is usually used instead when unidirectional change assumes relatively unpredictable forms; the future "evolution" of human society is much harder to predict than is the future "development" of a human infant. Since predictable and planned changes generally take place within the context of a wider range of unplanned and unpredictable change, the concept of evolution is more fundamental than that of development.

Social-Evolutionary Approaches

The analogy between societies and individual biological organisms, or the "organic analogy," has two implications that may be noted here.

(1) It implies that a society, like an individual organism, passes through a fixed sequence of stages such as infancy, childhood, adolescence, adulthood, and old age. This aspect of the organic analogy is illustrated by Arnold J. Toynbee's description of our own Western civilization, which, he said, developed within the womb of ancient Greece and Rome, was born in the Dark Age following the collapse of the Roman Empire, experienced its childhood in the Middle Ages, and attained the prime of adulthood as the great era of global exploration dawned around A.D. 1500.[2] (2) The organic analogy also implies that birth and growth are inevitably followed by decline and death, a point that Oswald Spengler stressed in his *Decline of the West*.[3]

However, instead of thinking of society as similar to an individual passing through stages in the life cycle, we might think of it as similar to an evolving species. This analogy suggests quite different features: a species, unlike an individual, does not progress through fixed stages and does not inevitably "die" at the end of a relatively fixed life-span; it *might* "die" (become extinct); or it might lose its identity in a process of change (i.e., evolve into another species) without actually dying; or it might survive for an indefinite time.[4]

Because basic stages in the life-span of a normal individual human being are obvious to people everywhere and at all times, the analogy between the human individual and the human society suggested itself in antiquity. Thus, the Roman writer Florus described stages in the life cycle of the Roman Empire, which he considered to be in its "old age" when he wrote in the first century A.D.[5] The evolution of species is not nearly as obvious as the development of individuals and has thus not been available until comparatively recently to serve as a basis for analogies with the evolution of human societies. In fact, the concept of evolution of species as an open-ended process rather than one that moves inevitably through fixed stages toward a predetermined final state was not clearly formulated until the appearance of Charles Darwin's *The Origin of Species* in 1859. Theories of social evolution that suggest at least a partial analogy with the biological evolution of species had appeared well before that time.

However, the contrast between theories of societal development based on the organic analogy and theories of societal evolution based on a species-evolution analogy is not as clearcut as the preceding remarks might suggest. Organic-analogy theorists in the nineteenth and twentieth centuries have often cautiously noted that the analogy is not perfect and that the development of society is less predictable, following a less rigidly determined pattern, than the development of the individual person. Toynbee, for example, uses the organic analogy, but also emphasizes that societies differ from individual persons in that they do not have fixed life spans; a society—or "civilization," to use Toynbee's favored term—may endure for thousands of years after attaining the equivalent of "old age" *or* may disintegrate rapidly. On the other hand, in the eighteenth and nineteenth centuries, several theories of societal change were formulated that were based more on the evolutionary than the organic analogy in that they postulated no ultimate deterioration or death but rather continued growth or progress. These theories nevertheless retained certain traces of the organic analogy in that they envisioned a fixed sequence of developmental stages rather than an open-ended evolutionary process.

Some of these theories focused on the evolution of *ideas,* with the assumption that ideas were the decisive element in human progress: for example, the Law of the Three Stages, formulated by Auguste Comte (1798–1857), according to which mankind moved from "theological" to "metaphysical" and ultimately to "positive" (scientific) thinking. A quite different strand of social-evolutionary theory focused on social organization and technology, for example, the scheme of Marie Jean de Condorcet (1743–94), who traced human development through nine stages beginning with the primitive "hunting and fishing" horde and continuing up to the French Revolution; and that of Adam Ferguson (1723–1816), who described "savagery" and "barbarism" as stages in the progress toward "civil society."

A major theory emerging from this latter approach was presented by Lewis H. Morgan (1818–81) in 1877. According to Morgan, evolving societies pass through sequential stages including lower, middle, and upper stages of savagery; lower, middle, and upper stages of barbarism; and ancient and modern stages of civilization. These stages are defined primarily in terms of material–technological criteria: the middle status of savagery commences with fish subsistence and the use of fire, the upper status of savagery with the bow and arrow, the lower status of barbarism with use of pottery, the middle status of barbarism with domestication of animals (in the Eastern Hemisphere) and with use of irrigation and adobe-brick and stone houses (in the Western Hemisphere), the upper status of barbarism with manufacture of iron, and ancient civilization with the phonetic alphabet and literary composition in writing. Morgan claimed that each of his major evolutionary stages is associated with characteristic forms of family and of economic and political organization.[6]

Morgan's scheme, published more than a century ago, is inevitably inadequate in certain respects, just as other older schemes are. Morgan made various factual errors and also erred by assuming that evolutionary stages were fundamentally similar for all societies. Today we understand the need for more flexibility than Morgan showed: the need to allow, more than he allowed, for variations among the societies at any given stage, for societies that "skip" stages, and for societies that combine diverse characteristics in ways that make them hard to classify in terms of "stages."

In 1884 Friedrich Engels (1820–95) incorporated Morgan's ideas into the Marxist scheme that he and Karl Marx (1818–83) had jointly developed. Engels also elaborated upon Morgan's last stage, that of "civilization," describing civilization as passing through the stages (or substages) of slavery, feudalism, capitalism, and (in the future) communism.[7] Marx and Engels viewed social evolution as a dialectical process in which each stage (prior to the final stage of communism) contains seeds of its own destruction in the form of internal "contradictions" that trigger the transition to the next stage. Thus, "contradictions" within capitalism (e.g., greatly increased productive capacity combined with low purchasing power that prevents consumers from buying the goods thereby produced) stimulate movement toward socialism, which is a transitional phase between capitalism and communism. Marx and Engels, along with Morgan, assume an initial condition of egalitarianism and communism in primitive societies, which gives way to more complex social organizations in which inequality and exploitation prevail, to be followed in turn by a movement back toward egalitarianism and communism, but at a higher level, in a much larger and more advanced society. In a famous assertion, which Engels cited with strong approval, Morgan boldly prophesied that "the next higher plane of society to which experience, intelligence and knowledge are steadily tending . . . will be a revival, in a higher form, of the liberty, equality and fraternity of the ancient gentes."[8]

The Marx-Engels scheme, although still alive today in various updated versions (one of which is incorporated into the official ideology of the Soviet Union), nevertheless has features that serve to remind us of its nineteenth-century origins. This scheme posits a fairly rigid sequence of stages culminating in a final, permanent condition of "communism," which the scheme describes only in an abstract, hypothetical way because no society has ever reached this condition. Furthermore, this final condition is a happy one and resembles in certain respects the condition that prevailed initially, before the movement through the sequence of stages began (i.e., the "communism" that represents the final outcome of the evolutionary process is merely a higher-level counterpart of the "primitive communism" that characterized all humanity in the beginning). The fixed sequence of stages in

Marxist theory suggests an implicit reliance on the organic analogy. The absence of any final deterioration of "death" of society at the end of the postulated sequence of changes suggests a biological-evolution analogy. But in this case the most relevant analogy may be with Christian theological doctrines that envision an initially happy state (Adam and Eve in the Garden of Eden), a series of transformations (beginning with expulsion from the Garden), and a final happy state again (salvation for the faithful).[9]

In the late nineteenth and early twentieth centuries, several investigators described a major transition that appeared to be taking place, from an older traditional type of society to a newer modern type. Each of them used somewhat different concepts to describe this transition and to contrast the new with the old, and each focused on a different aspect of the transition and the contrast. Hence we find, in the sociological literature of that time, what have come to be interpreted as basically dichotomous distinctions between *Gemeinschaft* (community) and *Gesellschaft* (society),[10] between status and contract,[11] between sacred and secular society,[12] between folk and urban society,[13] and between cohesion based on similarities among people ("mechanical solidarity") and cohesion based on interdependence among specializations ("organic solidarity").[14]

Also in the late nineteenth century, Herbert Spencer popularized a conception of evolution (including societal evolution as one of its many manifestations) as a process of differentiation involving a transition from simple to complex, or from homogeneous to heterogeneous.[15] More recently, Talcott Parsons has formulated a social-evolutionary theory in which patterns of cultural differentiation are employed as primary criteria in identifying evolutionary stages. "Primitive" societies are preliterate and thus lack cultural features complex enough to require writing. "Intermediate" societies do have written language but not the *mass* literacy with which we moderns are familiar. "Archaic" societies represent an initial intermediate stage: they have craft literacy and cosmological religion. "Advanced intermediate" societies have full upper-class literacy, and historic religions characterized by a differentiation between the natural order and a supernatural one. Because ideas put into writing can be stored and transmitted relatively easily from one social setting to another, the growth of written language and literacy has the effect of giving culture (or at least those aspects of culture that can be put into writing) a greater degree of autonomy relative to society than was possible before. Modern society extends literacy still further and is characterized especially by a "generalized legal order." Parsons's emphasis on cultural differentiation leads him to note also the existence and importance of tiny "seedbed" societies (ancient Israel and ancient Greece) that serve as incubation places for ideas subsequently applied on a large scale elsewhere.[16]

Leonard T. Hobhouse used intellectual and organizational criteria in classifying literate societies. However, insufficient evidence about the intellectual and organizational features of many primitive societies led him to the conclusion that it would be necessary to use material-technological criteria for classifying societies at *pre*literate levels.[17] The result is a pattern similar, in this respect, to the pattern we obtain when we combine the societal classifications of Marx and Engels with those of Morgan, as Engels (writing after Marx's death) intended them to be combined. Morgan, focusing on earlier societies, used primarily material-technological criteria of classification, whereas Marx and Engels, focusing on more advanced societies, recognized the importance of material technology in societal evolution but used classificatory criteria centering around socioeconomic organization and especially the organization and control of the means of production.

Some other twentieth-century investigators have focused more consistently on technological criteria. Lewis Mumford has distinguished among "eotechnic" (water and wood), "paleotechnic" (coal and iron), and "neotechnic" (electricity and alloy) phases.[18] Gerhard Lenski and Jean Lenski have defined stages of progress in terms of material technologies related to subsistence. In doing so they have adopted, with respect to the more primitive societies, concepts quite similar to Hobhouse's, but unlike Hobhouse they continue to use subsistence-technology criteria for classifying societies at *all* developmental levels.

The Lenskis recognize the following types of society to be in the main sequence of evolutionary development: "simple hunting and gathering" (more than thirty-five thousand years ago, with no tools more sophisticated than a wooden spear), "advanced hunting and gathering" (with spear-throwers and bow and arrow), "simple horticultural" (with plant cultivation but no metallurgy or plowing), "advanced horticultural" (with metal tools and weapons), "simple agrarian" (with agriculture and the plow), "advanced agrarian" (plow agriculture plus manufacture of iron tools), and "industrial." In addition, the Lenskis recognize several types of societies—fishing, maritime, herding—that occupy specialized niches and lie outside the main evolutionary sequence. Finally, "hybrid" societies combine features of two or more types.[19]

The Lenskis' approach, which assumes that capitalism and socialism are alternative ways in which an industrial society may be organized, contrasts strongly with the Marxist conception of societies as evolving from capitalism through socialism toward communism. Paradoxically, however, the Lenskis agree with Marx, Engels, and Morgan in assuming that social inequality and exploitation, which were minimal in the simplest societies and grew as preindustrial societies became more complex, tend to decline again as society continues to evolve (even though there is not much agreement between the Lenskis and Marxists as to what society is evolving toward in other respects).

Some doubts seem appropriate concerning *both* the Marxists' and the Lenskis' conceptions of movement toward greater equality. Marxists see such a movement with the emergence of communism, but communism remains today as a hypothetical condition not yet achieved anywhere; and when prestige and power as well as economic differences are taken into account, there is no basis for concluding that "socialist" societies, allegedly on their way toward communism, are any more egalitarian than contemporary capitalist societies are.[20] The Lenskis see movement toward greater equality with the development of industrialization, and this is a reasonable conclusion provided one compares industrial nations with the agrarian societies of earlier times. However, one might also reasonably assume that earlier agrarian societies should be compared not with contemporary industrial nations but with the contemporary world as a whole, and *this* comparison does not appear to provide any basis for recognizing an egalitarian trend.[21]

If the various societies of the world had no contact with each other, so that each one had to progress (or stagnate) independently of the others, *and* if all independently developing societies had to go through the same developmental stages (as theorists prior to the twentieth century commonly assumed), *then* one could reasonably view societal development as a process in which each society moved along the same road in the same direction, each at a different speed depending on its own unique circumstances. However, neither of these conditions actually prevails. First, societies do interact with, and influence, each other. Especially relevant here is the fact that comparatively advanced societies influence comparatively backward ones (and this is true regardless of the precise criteria by which advancement and backwardness are defined.) This phenomenon was often grossly underestimated in earlier times, as suggested by Ferguson's comment, made in 1767, that "if nations actually borrow from their neighbors, they probably borrow only what they are nearly in a condition to have invented themselves."[22] Second, there may be more than one path of development. These considerations open up a variety of possibilities.

1. The effect of contact between advanced and backward societies on the development of the backward societies may be either positive or negative (or a mixture of both, but to clarify relevant issues without becoming bogged down in details, we may best focus on the extreme alternatives). Backward societies may be able to take advantage of the experiences that advanced societies have already gone through, and may acquire, by imitation or importation, ideas and artifacts that the advanced societies have already created, discovered, or invented. They may thereby be able to *skip* certain stages of development. This possibility is especially clear when advancement is defined in material-technological terms. A society that has never had any airplanes and seeks to develop aviation today does not need to begin with the Wright brothers' model and work its way gradually toward more recent models but may instead simply import models that are comparatively advanced and skip the earlier ones. On the other hand, a drastically different consideration must also be taken into account: advanced societies may impede, rather than facilitate, the progress of societies that are more backward. In particular, an advanced society might monopolize scarce resources or opportunities and thus make it harder (or in an extreme case, impossible) for other societies to follow in its footsteps, or it may intervene to prevent other societies from doing so.

2. If a backward society is able to profit from the experiences of more advanced societies by skipping stages of development, there is a particularly interesting special possibility: the backward society, instead of (or in addition to) skipping stages that advanced societies passed through some time ago, might actually skip the stage in which advanced societies are *presently* located. Newly emerging technological ideas may be hard for an advanced society to put into practice because the advanced society has already acquired a commitment to an older system with which the new ideas would compete. But a more backward society, which never acquired the older system, has more flexibility in this respect. There are many possible examples, but perhaps the following will suffice. In the late eighteenth and early nineteenth centuries, Britain became the first society to industrialize. As improved kinds of industrialization were subsequently invented, Britain was slow to adopt them because it had an investment in the industrial equipment and procedures that it had acquired earlier. This made it easier for more up-to-date industrial practices to become established in Germany and the United States, which had not industrialized when Britain did, but which accordingly bypassed Britain to become the leading countries in the "second" Industrial Revolution of the late nineteenth century.[23]

3. If a backward society's development is blocked rather than

stimulated by contacts with more advanced societies, a distinction may also be made between two possibilities that then become relevant. The backward society may tend to remain in its original backward condition, *or* it may be transformed in a way that involves *not* "advancement" but rather the acceptance of a specialized subordinate role in a global system of societies that the advanced societies dominate. This latter possibility has been explored by "world system" theorists, who point out, for example, that some countries with colonial histories were compelled or encouraged to specialize in providing raw materials to advanced countries and to acquire technologies appropriate to this subordinate role, for example, domestic transportation systems oriented toward efficient exporting of raw materials rather than toward other purposes that might have been more consistent with local societal advancement.[24]

One cannot help but notice that several theorists have happened to find in their own respective countries the closest existing approximations to whatever kind of society they consider to represent the ultimate outcome of societal evolution: for example, the evolutionary stages formulated by the Frenchman Condorcet, culminating in the French Revolution; Soviet Marxists presenting the Soviet Union as leading the rest of the world as it passes through the stage of "mature socialism" en route to communism; and the suggestion by the American sociologist Parsons that "the United States is a model for other countries in structural innovations central to modern societal development."[25] Evolutionary formulations may indeed be profoundly influenced by the author's own personal identity. This is a common phenomenon only because societal evolution has no clear-cut pattern that is immediately obvious to all disinterested observers, but rather involves complex situations and trends that can reasonably be interpreted in diverse ways.

Several traditions of social-evolutionary theory remain alive today, including a Marxist tradition in several versions emphasizing social-organizational distinctions and especially the organization and control of the means of production; a tradition emphasizing cultural development and differentiation as exemplified by Parsons; and a tradition focusing on subsistence technology as in the Lenskis' formulation. In addition, there are diverse views concerning the processes by which societal evolution proceeds, and concerning the statuses of "hybrid" societies which exist by virtue of intersocietal contacts. I hope that this discussion has helped to clarify some of the issues involved.

The Evolution of Modern Society

Societies will be classified here according to the type of technology that predominates. This is justifiable on purely practical grounds. It is much easier to tell what kind of technology a society has or has had, than to classify a society in various other ways. This is true especially with respect to societies of the distant past, whose surviving artifacts tell us more about material technology than about values and organizational features. It is also true with respect to contemporary societies in many cases: for example, it is easier to describe the status of the Soviet Union with respect to technological development than to say whether or not the Soviet Union is a genuinely "socialist" society.

Technological criteria for classifying societies are also justifiable on the ground that technology has a special importance as a source and stimulus of societal evolution. This is paradoxical because evolution is an accidental, unplanned process whereas technology, by definition, is a planned phenomenon organized around means-ends relationships. The paradox can be resolved if we recognize that (1) humanity is disunited, with competing plans and hence uncertain outcomes; (2) planning something does not mean that we can control or predict or plan its *consequences;* and (3) our technology is imperfect, and plans will sometimes fail.

Technologies will be roughly divided into "premodern" (traditional) and "modern" types, with modern technology being that which is based on systematically organized innovation, whereas premodern technology emerged in more accidental ways.

Societies will correspondingly be classified into premodern and modern types, with allowance for intermediate cases and with one important qualification. A third general type of society will be recognized, including those that remained "behind" while other societies modernized and that are now in the process of "catching up." These societies, which Lenski described as "hybrid" and which will be called "transitional" here, have one central feature: because societies that are now modernizing can skip stages through which the first modernizers passed and can import the newest technologies to combine with much older technological forms already present, they tend to be characterized by extreme contrasts between the old and the new. Chapter 5 will be devoted to a detailed description of the technological characteristics of a single transitional society, the People's Republic of China.

Premodern societies—which have almost all disappeared or ceased to be premodern and should thus be referred to in the past tense—varied considerably among themselves in two quite different ways. First, there was an enormous range of variation between small primitive societies with relatively simple cultures and organizational arrangements and only a few hundred members or even fewer and other societies that, although also clearly premodern, nevertheless contained millions of people and produced such phenomena as irrigation systems, pyramids, huge marble buildings, cities, ships rowed by hundreds of oarsmen, complex systems of writing, mathematical theorems, and reasonably accurate predictions of some solar eclipses years in advance. Second, *within* each level or stage of societal evolution, there was a high degree of variation with respect to numerous aspects of culture and social organization. Although all *modern* societies today are interconnected and highly similar to each other in many ways by virtue of their interconnections, the numerous premodern societies of earlier times were relatively isolated from each other; each of them, evolving in its own local environment, acquired its own highly distinctive way of life.

Societies of the type we call "modern" first emerged within the framework of western European civilization in a protracted sequence of events beginning about five centuries ago. European explorers—venturing across the Atlantic, around the southern tip of South America into the Pacific, and also down the west coast of Africa and into the Indian Ocean—initiated a movement that ultimately made western Europe the first place in the history of the world to acquire contact with all the inhabited regions of the globe.[26] Major improvements in navigation and the growing use of firearms stimulated subsequent phases of western European expansion, which involved the subjugation of non-European peoples in large parts of the Americas, Africa, and Asia; the settlement of large numbers of western Europeans in new European-type nations abroad (most notably the United States); and a spreading of elements of the culture, including the technology, of western Europe, to non-European societies elsewhere.[27]

This geographical expansion of western European or "Western" civilization, which has brought all segments of humanity into contact with each other for the first time, has been associated with equally unprecedented and momentous developments within Western civilization itself: the Scientific Revolution of the sixteenth and especially the seventeenth centuries; the Industrial Revolution of the eighteenth century; and scientific and technological progress that continues to take place rapidly today and that has brought about an immense transformation in the conditions of human life.

Technology, as explained in Chapter 2, is as old as the human species, and older (with some technological phenomena discernible on a small scale among nonhuman mammals). Even the most primitive human societies have had technologies that have enabled them to accomplish such tasks as acquiring food, obtaining protection from the elements and from dangerous animals and human enemies, caring for young children, and coping with illness. Science, defined as a socially organized search for laws of nature, is not nearly that old. It is true that essential elements of "scientific method" were implicitly understood by a few thinkers in primitive societies and used by them in coping with some practical problems. Thus, the preliterate Seneca Indian leader Handsome Lake (1735–1850) proposed what amounted to a scientifically controlled experiment, although of course he did not call it that: he sought to compare the behavior of people who drank alcohol with the behavior of people similar in other respects who abstained.[28] But occasional instances of scientific-type thinking and observation in dealing with practical problems, by a few exceptional individuals in a society, does not mean that the society in question has acquired *science*.

In fact, science as we understand it exists only to the extent that (1) the scientific method of inquiry is employed for scientific purposes, that is, in pursuit of general knowledge of nature, rather than merely in pursuit of solutions to practical problems, and (2) a community of scientists has emerged, whose members share their observations and theories; critically evaluate each other's contributions; and accumulate, store, and transmit from one generation to another the body of knowledge thus acquired.

These conditions were first met, on a small scale, not in primitive societies, but in the larger and more complex civilizations of the ancient world. What we call *modern* science did not emerge until the Scientific Revolution of seventeenth-century western Europe. Its emergence involved several developments: (1) Science at this time made spectacular *substantive* advances, most notably with the appearance of Newtonian mechanics, which brought together in a single framework the analysis of terrestrial gravitation and the analysis of motions of the stars and which constituted a model for further scientific endeavors in diverse fields. (2) Science also made enormous *methodological* advances, involving a clarification of the procedures of controlled observation and experimentation and the procedures of deductive theory-construction. (3) Relatively stable and effective scientific communities concerned with wide ranges of natural phenomena appeared on the scene for the first time. (4) Science acquired self-reinforcing tendencies that have produced a spectacular growth of scientific knowledge continuing to our own time.[29]

The rise of modern science has been explained in a variety of ways. Theories on this topic tend to be highly controversial. I shall not discuss this question here except to point out that several technological inventions and the explorations mentioned above played a part. The appearance of instruments such as telescopes, microscopes, barometers, and pendulum clocks was a contributing factor. Galileo's discoveries of mountains on the moon, phases of Venus, sunspots, and satellites circling Jupiter, in the early 1600s, were consequences of his use of the newly invented telescope. And new information about strange lands and peoples, pouring into western Europe from the voyages of exploration, stimulated new ideas and thus helped to create a favorable milieu for scientific inquiry.

The Scientific Revolution was an international development within western European civilization, with major contributions, for example, from the Pole Nicholas Copernicus (1473–1543), the Dane Tycho Brahe (1546–1601), the German Johannes Kepler (1571–1630), the Italian Galileo Galilei (1564–1642), and the Englishman Sir Isaac Newton (1642–1727). The Industrial Revolution, which followed, was initially much more concentrated in one country, England, which made the huge transition to an industrial way of life between approximately 1760 and 1850. During this time the burning of wood as fuel largely gave way to the burning of coal; water and wind as sources of power were largely displaced by the steam engine; production and use of iron rose rapidly; new machines dramatically increased productivity in the textile industry; large factories in large cities drew significant numbers of workers from agricultural occupations in the countryside; and dramatic advances in transportation were symbolized by the steamship and by the railway train, which Lewis H. Morgan in 1877 called "the triumph of civilization."[30]

In its first (English) phase, the Industrial Revolution involved the practical application of the general concept of scientific method but not, to a major extent, a practical application of *particular* scientific discoveries. The latter came only with the Industrial Revolution's second phase, which was geographically more dispersed than the first phase but with some concentration in Germany and the United States. In this second phase, scientific knowledge in fields such as chemistry and electricity was applied systematically to practical use, for example, in the manufacture of aspirins and dyes from coal tar, which was a notable achievement of German chemists, and in the inventions of Thomas Edison in America.

I have mentioned earlier that technology is much older than science and that some technologies available in seventeenth-century western Europe helped to bring modern science into existence. New technology continues to play an important part in facilitating scientific inquiry today. However, with recent phases of the Industrial Revolution, the relationship between science and practical technology has become increasingly two-directional, with technological progress not only stimulating science but also coming to constitute largely an application of previously acquired scientific knowledge. This is illustrated most clearly in the development of the atomic bomb during World War II; this technological innovation was a direct practical application of new scientific knowledge in nuclear physics.

The next phase in human societal evolution cannot be predicted with confidence. There are numerous drastically divergent possibilities: total extermination of civilization or even of all humanity through warfare with new weapons of mass destruction; "nuclear blackmail" by terrorist groups armed with mass-destruction weapons, which could hold entire cities or countries as "hostages" and undermine the basic structure of the contemporary nation-state; collapse of modern civilization through damage done by overpopulation and/or modern technology to the natural environment on which civilization depends; global totalitarianism supported by new technologies of communication and control; the enslavement of humans by computers; a large-scale repudiation of modern science and science-based technology and reversion to simpler and more socially cohesive lifestyles; global communism with a withering away of the state as envisioned in Marxist doctrine; or a "postindustrial" or "postmodern" society growing out of contemporary modernism in its capitalist (especially American) form and out of an expansion of contemporary science and technology. Although the uncertainties of today's world give us no firm basis for knowing which of these (or other) possible outcomes will materialize, we may be reasonably confident that the *process* of societal evolution (assuming that there remains a society to evolve) will be generally consistent in its basic structure with the social-evolutionary process that has brought us to our present condition from primitive beginnings.

4
Directions of Technological Change

Premodern Technology

Let us begin with a brief analysis of the traditional subsistence technology of a small, primitive, hunting society located in an exceedingly challenging natural environment—as this technology existed in the old days prior to major contact with modern civilization.

The traditional subsistence technology of the Netsilik Eskimos of north-central Canada includes kayaks; waterproof and easily transportable sealskin tents; igloos that took advantage of the prevalence of snow in winter and of snow's insulating properties; sleds designed so they could be dragged over rough ice without breaking; domestication of dogs and their use for many purposes including the pulling of sleds; caribou-skin and seal-skin clothing; diverse kinds of harpoons, spears, and knives, and bow and arrow; fishing and seal-catching through ice holes in winter; and several methods of caribou hunting.[1]

This technology was linked with two other features of the Eskimo society. First, there were rituals associated with it that were thought to facilitate success in relevant endeavors: thus, the souls of killed seals had to be treated with respect in order that seals encountered in future hunts would be willing to be killed. Second, the social organization of the Eskimos, in particular their family organization, was patterned in a way that encouraged cooperation of the type that was essential if the subsistence-technology system was to operate effectively. Details concerning the rituals and the family organization are not relevant here. These aspects of the Eskimo way of life are mentioned here only to point out that neither of them is "technological" and to clarify still further the meaning of "technology" in the present context. The rituals were indeed intended as "means" for achieving "ends" and thus satisfied *one* of the requirements of our definition of technology, but they were not genuinely technological because the means involved were supernatural rather than natural. The family organization, on the other hand, apparently did contribute in a wholly "natural" way to success of the subsistence technology; but it presumably represented the outcome of an unplanned evolutionary process maintained by traditional values rather than a means deliberately instituted and maintained in pursuit of any end, and thus failed to satisfy a different requirement that our definition of technology entails.

The subsistence technology of the Eskimos was excellently adapted to the local natural environment. This means not only that there was a close integration between this technology and its environment, but also that this integration was achieved primarily by bringing the technology into line with the environment rather than the environment into line with the technology. Such a pattern is characteristic of stable primitive as distinct from more advanced technologies: thus the American Indians of the Northeast woodlands acquired a technology appropriate for woodland life, whereas more modern white invaders from Europe used their technology in a way that largely destroyed the woodlands.

However, certain qualifications and clarifications need to be made to the statement that primitive technologies tended to adapt to preexisting natural environments:

1. The statement applies only to primitive societies living under comparatively stable conditions prior to the recent intrusions of modern civilization into the lives of primitive peoples everywhere. It does not apply to peoples whose primitive way of life is in the process of destruction through modern influence or has already been destroyed; in fact, such people are no longer truly "primitive."

2. The statement should not be interpreted to mean that primitive societies have *no* environmental impact. No species in an ecosystem fails to affect the functioning of that ecosystem in some way. We should hardly expect even the most primitive humans to constitute an exception in this respect. However, in the case of a typical primitive society, the societal impact on the ecosystem is (a) quite limited in scope and intensity by comparison with the impacts of more advanced human societies, and (b) relatively constant and consistent with ecological stability, rather than rapidly accelerating in a way that is incompatible with stable ecological relationships.

3. The statement should not be interpreted to mean that the

natural environment "determined" or "largely determined" the character of primitive technology. It *does* mean that the natural environment imposed certain limitations on the range of variation in subsistence technology that was feasible; but within the range that was *not* thereby precluded, other factors may have had important and even decisive influence. Thus, a tropical environment precludes an Eskimo-type subsistence technology; but there may nevertheless remain a variety of quite different potential primitive subsistence technologies, all within the range that a tropical environment permits. A tropically based primitive society's selection from among these alternatives was not always necessarily strongly influenced by environmental considerations. On the other hand, *some* natural environments—very likely including the Eskimos'—are highly restrictive, not permitting much "choice" among primitive subsistence technologies.

4. One additional clarification must be made here. I am *not* saying that primitive peoples have greater "respect" for the natural environment than modern peoples do. This may indeed be true in some cases; but it is not necessarily, and perhaps not usually, true. Nor am I denying that primitive peoples commonly engage in behavior that *would be* highly destructive of existing natural environments if it were manifested on a greater scale. A purely imaginary example will illustrate the point. Primitive people living in a forest might be so *disrespectful* of their environment that they actually undertake to destroy it by chopping down every tree. But if the forest is large, the people are few in number, the trees are sturdy, and the tools for tree chopping are inefficient, the forest might remain essentially undisturbed regardless of the "disrespect" that its human inhabitants display toward it. To consider a more likely example, a primitive group may engage in hunting and seriously deplete the supply of game in the immediate area. But if the group has a large territory available, it may move from one part of that territory to another part every few years, so that small local areas within that territory that have been overhunted have an opportunity to recover. Given a small enough group of people, a large enough territory, comparatively weak technological capacities and environmental damage limited to types from which recovery in a few years is possible, such an arrangement might continue indefinitely with no long-range environmental damage to the territory as a whole.[2]

In saying that a technology is adapted to the natural environment, we are thus not implying anything about the "attitudes" of the people who use the technology. It is entirely possible that a technology may be fully consistent with environmental stability even if attitudes of those who use the technology are disrespectful, provided the characteristics of the technology and of the environment, and the relation between population and resources, make it impossible for that disrespect to produce actual environmental damage or other change. Furthermore, if any damage that does occur is localized, a roving way of life (or other adjustments) may sometimes permit recovery and thus permit the technology that does the damage to be consistent with long-range stability in the environment as a whole.

Conversely, when primitive people do show real respect for the environment or for certain aspects of it, this does not necessarily have any objectively beneficial environmental effects. Some primitive peoples, including Eskimos, routinely engage in certain rituals that show respect for the animals they kill. However, such rituals, even if they constitute a demonstration of "respect" for the environment or for certain environmental components, do not necessarily have any beneficial or preservative effects on the environment: one may "respect" the environment while destroying it.

Primitive technologies tend to be well integrated not only with the natural environment but also with other aspects of the societies in which they are embedded. In particular, primitive technologies tend to be consistent with (1) the requirements for societal survival under stable primitive conditions; (2) motivations, skills, and resources among the societal population, which are necessary to keep the technology functioning; and (3) the prevailing culture and social organization. Thus, a stable primitive technology typically provides food and protective clothing and shelter; does not make insupportable demands on motives or skills of the people or on available resources; and fits in nicely with the religion, family structure, and other cultural and organizational features of the local society.

The subsistence technology of the Netsilik Eskimos was drastically limited in power by its heavy reliance on human and animal muscles. This technology was also relatively simple, although not quite as simple as a subsistence technology among primitives could conceivably be. It was complex enough to entail skills that could only be acquired through prolonged training and experience, and complex enough to involve differentiation of tools for similar but distinct purposes (e.g., different types of harpoons or spears for hunting different types of animals). On the other hand, there was *no* systematic division of labor involving differentiation of technical skill except on the basis of gender: any man (or woman) could normally do what any other man (or woman) could do. Simplicity in the technology-related division of labor was also facilitated by the reliance on materials available locally (which eliminated division-of-labor complexities associated with acquisition of materials from a distance) and by the small number of people involved in the society and hence in the technology (which eliminated division-of-labor complexities associated with management and coordination of large work forces). Another relevant indicator of technological complexity is the use of secondary tools (tools used to make other tools), tertiary tools, and so forth; the Netsilik appear to have had a subsistence technology that was relatively simple in terms of this criterion.[3]

For further clarification of the characteristics of premodern technologies, it will be helpful to compare primitive technologies such as those of the Eskimos with other technologies that are as far removed as possible from the primitive level while also clearly within the "premodern" category. An outstanding example of a highly advanced premodern technological achievement is the Great (Khufu) Pyramid of Egypt. This pyramid is particularly noteworthy because it combines extremely great antiquity (as compared with most other extant architectural works of ancient times) *and* features of scale and quality that remain impressive today.

Constructed around 2600 B.C. (and thus already, at the time of the Roman Empire's founding, several centuries older than the Roman Empire would be today), this pyramid illustrates ways in which the material technology of comparatively complex premodern societies far transcended the primitive level. Built with approximately 2,300,000 stones, weighing an average of 2.5 tons each, the Great Pyramid and its two neighboring pyramids contain (according to calculations made by Napoleon nearly two centuries ago) enough material to build a wall three meters high and one meter thick completely surrounding France. Originally 482 feet high, the pyramid has a base covering thirteen acres. It is oriented so that its sides point due north, south, east, and west respectively, and the difference in length between the longest and the shortest side is less than eight inches. Inside the pyramid is a network of passages and chambers. The ancient Greek historian Herodotus, who visited the pyramid around 440 B.C. (when it was already about 2160 years old) reported that according to legend its construction required one hundred thousand workers and twenty years. These figures have been viewed skeptically by some recent investigators; but there is no doubt that the Great Pyramid, along with other pyramids of ancient Egypt, was a massive undertaking that required a huge investment of resources as well as remarkable engineering skills and impressive planning and coordination.

It is noteworthy not only that a project involving such resources and skills could have been planned, organized, and brought to fruition so long ago, so early in the history of human civilization, but also that this could be done even though the project contributed nothing, as far as we know, to subsistence or to societal defense against invaders or to solution of any problem that could be classified as "substantive" rather than "symbolic." In fact, the Great Pyramid was a pharoah's tomb, and to the best of our knowledge its function was symbolic only. There is no reason to deny or minimize the importance of symbolic functions, although in the absence of adequate written records we may not be able to tell in precise detail what these functions for an ancient technological item actually were. However, when a society can afford to divert resources away from subsistence-related activities to build the largest tomb that anyone has ever had, we have an indication that, in one way or another, a significant economic surplus was somehow acquired.

Some other large-scale engineering projects in ancient times were more clearly utilitarian. One of these, the Great Wall of China, which stretches for about fifteen hundred miles through varied terrain, served as a barrier against "barbarian" invaders. Although the Great Pyramid of Egypt was planned and constructed in a fairly compact interval of time, the Great Wall of China had a very different history. Different segments of it were constructed at different times by different local rulers to defend their own respective local territories; and the decision to link these different segments together into a single Great Wall was made (in the third century B.C.) only after local segments had already been constructed.

The best examples of ancient large-scale engineering projects directly related to subsistence were the irrigation systems constructed in several of the great river valleys in the ancient world: first, apparently, in Mesopotamia, and later in Egypt, China, and India. These illustrate particularly clearly the tendency, as societies moved from primitiveness to civilization, for technologies to *change* the natural environment to an increasing extent, as distinct from merely *adapting* to this environment. Ancient irrigation systems are also particularly interesting because they have become a major focus for studies concerning the relationship between technology and social organization. Karl A. Wittfogel, in an important and controversial work, suggested that the practical requirements for constructing and maintaining large irrigation systems encouraged highly centralized and autocratic governments in what he called "hydraulic" societies dominated by irrigation.[4]

Despite the enormous gap that separates the technological level of the great ancient civilizations that produced pyramids, irrigation systems, and the Great Wall from the technological level of Eskimos and other primitives, these levels may nevertheless be usefully grouped together, as "premodern" or "traditional," on the basis of certain shared features.

1. Traditional technology tended to be comparatively stable or stagnant. There were occasional brief episodes of rapid innovation in particular times and places with respect to particular kinds of technology, but these stand out as unusual against the background of general and prolonged stability. For example, the techniques involved in pyramid construction were apparently mostly acquired by the Egyptians within a remarkably brief 300 years. Many inventions that from our modern point of view seem "obvious" were very long delayed or even completely absent in certain major premodern societies: the great ancient civilizations of central and South America never invented the wheel; and in the Old World thousands of years elapsed *after* horses had been domesticated before they came to be equipped with effective saddles, stirrups, iron horseshoes, and harnesses that enabled them to pull heavy loads without choking.

2. Traditional material technology relied relatively heavily on human (and sometimes animal) muscles. Giant engineering projects such as the Great Pyramid and the Great Wall, as well as the much more modest efforts of primitives, were undertaken primarily with muscle power. And some of the most impressive tools and machines of ancient times were intended *not* as "labor-saving" devices but rather as ways in which muscle power could be more effectively harnessed. Thus, the huge oared ships of Mediterranean antiquity required hundreds (and in extreme cases, thousands) of oarsmen per ship. A major problem in designing these ships was to figure out how the greatest possible number of oarsmen could be used without their getting in each other's way.[5] Similarly, the implements used to destroy or breach the walls of a city during ancient military sieges were often extremely labor intensive: at the siege of Rhodes in 309 B.C., one "fighting tower" was moved up to the walls by thirty-four hundred men; and a battering ram 180 feet long was handled by one thousand men.[6]

3. Traditional technological knowledge, even when impressively advanced, tended to be derived from trial and error rather than from systematic scientific inquiry and to be employed without real understanding of the scientific principles involved. Thus, metals were extracted from oxide ores for millenia without any scientific knowledge of the chemical processes by which this result was obtained. Similarly, trial and error in ancient times led to construction of ships that were

proportioned appropriately to minimize water resistance while protecting stability, even though relevant scientific principles were not to be clarified until modern times.[7] And, it is most doubtful that there was much explicit awareness in ancient Egypt of the abstract scientific principles that the pyramid builders implicitly utilized.

In the 1930s the inhabitants of a Balinese village were found to be using an interesting method of teaching their babies that (in accordance with local custom) they should never reach for anything with the left hand: a baby's *right* hand was immobilized, so he could only reach with the forbidden left hand, and when he began to do so, his mother would pull the left hand back and release the right one.[8] I have asked many of my students what method they would use if they wanted to teach children too young to understand verbal instructions that they should never reach with the left hand. I have never yet received an answer that seems as good as the answer that the Balinese villagers have found. (Most students say, "Hold the left hand so the baby can only reach with the right.") This does not mean that the Balinese villagers are more intelligent than American students: they have found a better solution to the problem in question because they have had centuries to think about it and to experiment, whereas I usually give my students only two or three minutes. We should not underestimate the capacity of ordinary people to solve technical problems via relatively inefficient trial-and-error methods, if there are enough people and if they are given enough time.

But we should also be aware of the limitations of this phenomenon: it often yields only very slow results; it may leave striking gaps in technological development in areas that are not perceived as problematic and thus do not receive profound and prolonged attention; and it often yields only low-level, concrete, how-to-do-it knowledge (the Balinese mothers presumably do not understand the psychological principles underlying their method of teaching of children to reach with right hands exclusively, any more than ancient craftsmen extracting metals from ores knew what chemical principles they were employing.) Limitations such as these have been overcome with the breakthrough into the realm of systematically organized science, which has provided a basis for technologies of the sort we call "modern."

Aspects of Technological Modernization

The transition from the earliest to the most modern technologies has involved developments in several directions, which may be summarized as follows:

1. Human muscles have come to be supplemented by the use of material tools. The simplest of these are merely objects occurring in the natural environment that are used in the forms in which they naturally appear: for example, sticks and stones picked up and used as found. Material technology at this level occurs among prehuman animals as well as among humans. A step beyond this involves the preparation or manufacture of tools. In its earliest and most modest forms, this entails only minor modifications of natural materials and may also occur among animals, as when chimpanzees remove leaves and bent ends from sticks that they then use to catch termites.[9] Subsequent developments are found only at the human level: the retrieval of "pure" substances from natural compounds and mixtures (e.g., the extraction of metals from ores); the deliberate creation of artificial substances (pottery, paper, glass, bronze and other metallic alloys, concrete, etc.); the use of secondary tools (tools used to make other tools) and tertiary tools and so forth; and the emergence of machines with standardized components that partly replace, and partly extend the capacities of, human muscles and (recently with computers) the human brain.

2. Human muscles provided the earliest form of power in human society. From this beginning one line of development involved devices by which increasingly large numbers of people could combine the strengths of their muscles; for example, ancient oared ships each manned by hundreds of oarsmen, whose locations on the ship were planned so as to prevent them from getting in each other's way and to permit them to coordinate their rowing. Another line of development involved the harnessing of animals (e.g., horses pulling plows), a process that in ancient times was rendered highly inefficient by harnesses that fastened around the neck and would choke an animal that was given a heavy load to pull. Also in premodern times readily observable aspects of the natural environment came to constitute sources of power: wind, water, and wood. The "black rocks" that the thirteenth-century Italian adventurer Marco Polo found the Chinese burning for fuel indicate emergence of fossil fuels as a power source. The Industrial Revolution saw the large-scale development of the steam engine, electricity, and chemical fuels. In our own time, modern science has brought us power sources previously unknown, including nuclear energy.

3. Technological modernization has meant a long-range increase in the productivity of labor, that is, a typical worker has come to produce more under modern conditions than was possible before. (We are concerned here with long-range trends and not with short-run cyclical fluctuations.) In addition, new *kinds* of production have become possi-

ble, that is, workers now produce goods and services of types previously unknown. The trend toward increased productivity is somewhat obscured by another trend toward increased complexity in the division of labor. Thus, we do *not* have a transition from "one worker producing one item in one day" to "one worker producing *five* items in one day"; instead, we find a transition from "one worker producing one item in one day" to "one *hundred* workers producing *five hundred* items in one day" (with each of these hundred workers performing a specialized task and none of them producing any completed items on an individual basis). The unemployment that would otherwise result from increased productivity per worker has been countered, in the long run, by an increase in the quantity and variety of goods and services for which there is effective demand.

Although increased productivity of labor has been historically important in the emergence of modern society, it does not necessarily have the same significance for transitional societies now undergoing a "belated" modernization. A comparison of agriculture in the United States and in China will illustrate this point. Today one American farmer feeds fifty-six people and would be capable of feeding more if the most advanced technology were employed on *all* American farms and if relatively inefficient farms were to be upgraded or abandoned. One Chinese farmer feeds only three people. But although the United States is thus very far ahead of China in productivity *per farmer*, it has a much more modest lead over China in terms of productivity *per unit of land* (hectare or acre). American agriculture developed in the context of an abundance of land and a shortage of labor; hence increased productivity per farmer was an important consideration. As George Washington wrote in 1791, "The aim of the farmers in this country . . . is, not to make the most they can from the land, which is, or has been, cheap, but the most of the labour, which is dear."[10] Chinese agriculture, by contrast, has been plagued by a shortage of good agricultural land and by an *excess* of farm labor (except at a few crucial times of year; there are labor shortages, for example, briefly at harvest time). Efforts to improve Chinese agriculture have quite reasonably focused on increasing productivity per hectare rather than productivity per farmer. More generally, transitional societies with abundant labor but with scarcities of other resources might quite reasonably develop technologies that differ from those characteristic of earlier modernizers, emphasizing increased productivity with respect to those other resources rather than with respect to labor.

4. Still another direction in which modern technology has been moving is toward increasing complexity, as measured by criteria such

as amount of specialized training required to understand, invent, produce, operate, and maintain it; the division of labor thereby required; variety of tools and artifacts required; use of secondary tools (tools to make other tools), tertiary tools, and so forth; variety of raw materials utilized; and extent to which raw materials are transformed.

However, modernization has also involved some paradoxical movements at the same time toward greater simplicity in certain respects, as shown in the following examples.

The "automobile system," broadly defined, encompasses the extraction, production, and transportation of all materials used in manufacturing, servicing, and repairing a car (including parts of a car, gasoline, tools used in auto factories, etc.); the auto-manufacturing process; arrangements for advertising, selling, and insuring cars; legal arrangements pertaining to licensing of cars and drivers; traffic controls and enforcement of traffic laws. This system has become vastly more complex, although certain parts of it have become simpler: a 1982 model car is easier to drive than a 1932 model car was, and some operations in the auto-manufacturing process have been simplified to the extent that production workers can be replaced by robots.

There have been dramatic simplifications in medical treatments for certain diseases. Lewis Thomas has pointed out that a case of typhoid fever in 1935 entailed about fifty days in a hospital with careful monitoring and specialized care, whereas today the cost can be merely "a bottle of chloramphenicol and a day or two of fever."[11] However, the entire technology of medical care has become more complex, not simpler. As particular ailments become simpler to treat, other ailments that previously could not be treated at all acquire modes of treatment that are often initially quite complex.

In 1892 William Crookes foresaw the "bewildering possibility of telegraphy without wires, posts, cables, or any of our present costly appliances."[12] Wireless telegraphy, which came as Crookes predicted, did involve technological simplifications as he indicated; but our communications systems today are much more complex than they were when wireless telegraphy was only an unrealized dream.

Scientists strive for simplicity in their concepts and theories or for "parsimony", to use the term that they prefer. Given two competing theories that cover the same range of phenomena and that are equally adequate empirically, scientists tend to prefer the more parsimonious alternative, although this tendency is often partly obscured by the fact that two theories are rarely if ever identical in both empirical adequacy and in the range of phenomena covered. Occasionally a new theory will produce a major simplification of knowledge in a given field. At the same time, however, new complexities will be arising in other fields of science, including some that are newly opening, so that scientific knowledge as a whole becomes more complex as scientists strive to simplify it to the greatest extent that requirements of empirical adequacy will permit.

Modern machinery has become highly complex in the sense that it often has a large number of separate parts, but there has also been a major simplification through *standardization* of parts, so that defective or outworn components of a machine can be replaced with spares and with no significant departure from the original mode of functioning. In 1867 the Walter A. Wood Mowing and Reaping Machine Company of Hoosick Falls, New York, distributed advertising in which parts for mowers and reapers were identified by number, so that purchasers could merely specify the numbers of those parts that they needed.[13] Standardization of machine parts, of which this provides an early illustration, came to be paralleled by standardization in the effective functioning of officials in bureaucratic organizations.

The division of labor has become immensely more complex with modernization, in the sense that the number of distinct occupations is far greater than before. However, in certain specific areas, simplifications of the division of labor have also occurred. For example, a sharp separation and prestige differential between mental work and manual work was characteristic of premodern civilizations and has been declining with modernization. This distinction is illustrated by the medieval European practice of surgery (manual labor!) by *barbers* (who were manual laborers) rather than by physicians (who were not). It inhibited the development of science, which requires that abstract thinking and manual-labor experimentation be undertaken by the same people. Although the decline of this distinction constitutes a simplification of the division of labor, it also makes possible a vast increase in the complexity of the division of labor: specifically, it makes possible a huge variety of new occupations that cut across the old boundary between mental and manual work (e.g., the modern surgeon and modern scientists in numerous specializations).

Convulsive attacks on the distinction between mental and manual work were a central feature of Chinese Communist policy under Chairman Mao Zedong, especially during the Cultural Revolution of the late 1960s. These attacks involved, among other aspects, the sending of scientists and other intellectuals to do ordinary manual labor among the peasants, while the latter were encouraged to participate in scientific research. Elsewhere I have suggested that these

attacks "might well have been paving the way for an . . . increase in
the complexity of the division of labor in Chinese society" even though
the Chinese Communist leaders intended exactly the opposite: "They
sought to smash the distinction between intellectual and manual work
not in order that more complex distinctions could emerge in its place,
but rather as part of an effort to simplify the Chinese occupational
structure."[14]

5. Technological modernization has created new patterns of interdependence among different geographical regions. Thus, the American
economy today is dependent upon materials that must be imported
from distant lands (among these materials, oil and certain "strategic"
metals have attracted considerable attention in recent years). This
situation contrasts strikingly with that of the original inhabitants of
North America, whose economies were based exclusively on their own
continent's resources and largely on local resources. However, once
again we find important instances in which a general trend associated
with modernization has been partly reversed. Prior to the First World
War, Germany depended on nitrates imported from Chile, which were
used to produce fertilizer and explosives. The wartime blockade cut off
Germany from Chile, which stimulated the Germans to develop chemical substitutes for what they had imported; and the German dependence on Chilean nitrates ended. More than a century earlier, the
continental blockade during the Napoleonic Wars had stimulated the
replacement of imported cane sugar with locally produced beet sugar
in Germany.[15] In the United States there have been similar efforts in
recent years to develop substitute fuels that would reduce American
dependence on imported oil—just as Benjamin Franklin's stove was
intended to substitute American wood for imported coal two centuries
ago.[16]

6. Modern technology has also involved movement toward systems which are *self-correcting*. Three examples, each very different
from the others, will suggest the wide diversity of ways in which this
tendency is manifested:

—Science is a self-correcting process in the sense that it incorpo-
rates within itself the means by which errors can be detected and
weeded out; science in this respect is fundamentally different from
various older systems of thought.

—Modern machinery often contains self-correcting devices such
as thermostats, which keep temperatures within set limits "automatically" without direct human intervention.

—Modern governmental systems tend to contain self-correcting mechanisms: for example, arrangements for *automatic* reapportionment of representative legislative bodies on the basis of changing census data, and arrangements for orderly transfers of authority.[17] The *absence* of such a mechanism is illustrated by the 1969 Constitution of the Chinese Communist Party, which specified that the soon-to-be-disgraced Lin Biao was Chairman Mao's "close comrade in arms and worthy successor"; to imagine a comparable rigidity in America we would have to suppose that the name of George Washington's intended successor as President was written into the United States Constitution.

7. Traditional technologies were primarily oriented toward adapting to preexisting environments; and, because environments were highly varied, technologies were often highly localized. Modern technology tends to change the environment rather than to adapt to it (although the difference is only one of degree); and, because the changes thus brought about are generally similar everywhere, modern technology tends to be comparatively widely applicable. It is true that mistakes have often been made by those who seek to transfer the highly advanced technologies of the most modern societies to more backward societies without taking local conditions into account. There is a large body of literature filled with warnings to the effect that technologies suitable in the United States and the highly advanced countries of western Europe may not be suitable at all in a village in India or Mexico. Thus, "monocultural" corn production, with *no* crops planted between the rows of corn, may be suitable in Iowa where corn fields are large and tractors are used; but it is inappropriate in a Mexican community in which farming is done by cheap labor without tractors and in which beans or other crops planted between the rows of corn can increase per-acre yields.[18] But considerations of this sort, however important, should not blind us to the tendency for modern technology to have greater worldwide relevance than the technologies of particular traditional societies that are closely linked with highly localized cultures and environments.

8. The idea that modernization entails a shift from "accidental" or "chance-determined" change to deliberately planned change is commonly accepted. Thus, one sociologist writes that "chance has played a declining role in history, through the rise of scientific understanding and of deliberate . . . inventing."[19] However, this idea needs to be modified to allow for the following considerations:

—The amount of deliberate change *and* the amount of accidental change may *both* have been increasing, as part of a general increase in the total amount of change in society.

—The boundary between deliberate change and accidental change is not completely clear. Hitler deliberately planned the actions that initiated World War II in Europe, but accidental events played a large part in the process by which he acquired the power to implement his plans. And, deliberately planned systems may have accidental consequences or malfunctions. Thus, the deliberately constructed nuclear arsenals of the superpowers could lead to an outbreak of nuclear war through a purely accidental sequence of events, *without anyone* "planning" or "intending" this result. Similarly, deliberate control of the weather, when this becomes feasible, may encourage us to produce weather patterns that have unplanned, totally accidental consequences for agriculture or for other activities.

These complications mean that, in practice, the massive emergence of advanced modern technology is associated with predictive and control capacities that are highly developed within certain narrowly defined contexts but that remain extremely limited outside of these contexts: we can confidently predict the outcomes of many diseases and the times at which airplanes will reach their destinations, but the long-range survival of the human species is highly unpredictable.

9. Modern technology is dynamic, continually changing and advancing. Its dynamic character, which contrasts strikingly with the relatively static character of traditional technology, has several sources.

—Although new technology may sometimes provide a stable satisfaction of needs, it may also, on the other hand, create new needs that call for still more technology. This is especially clear in the military domain: weapons call for counter weapons that call for still newer weapons to counter *them,* and so on. Furthermore, any complex modern technological item will require a large body of "supporting" technology; to cite one of many possible examples, large-scale use of the automobile in societies without much domestic oil production made necessary the invention and use of very large tankers capable of transporting huge quantities of oil around the globe.

—New technology may not only increase the *need* for still newer technology; it may also increase the potential *opportunity* for the latter to emerge. The greater the number of technological items already present in a society, the greater will be the number of possible new combinations of these items; hence a tendency for technological inventions to accumulate in accordance with the compound-interest principle.[20]

—In modern times, increasingly effective new methods of invention have themselves been invented; these involve systematically organized research by scientific and technological specialists within laboratories and other organizations devoted at least in part to such research. *This* development is central to technological modernization: it is the feature in terms of which "modern" technology has here been defined.

Retention And Revival of Older Technological Forms

Chapter 2 pointed out that old established technologies sometimes lose their "technological" status and become instead merely part of the traditional cultural background of society. Old technologies may also disappear as new ones arise. Lewis Mumford has emphasized that this tendency may have extremely harmful consequences: when a body of highly diversified local techniques is displaced by a single standardized technological system, there is a potential loss in flexibility, with retreat to the earlier pattern no longer being possible because the older technologies are forgotten or because the conditions under which they flourished have been eliminated.[21]

On the other hand, the appearance of new technology does not by any means automatically eliminate the older technologies that it displaces (but that it often does not *completely* displace). Radio did not eliminate newspapers; television did not eliminate radio; automobiles did not eliminate bicycle riding or horseback riding; and such old-fashioned technological items as hand shovels, wheelbarrows, canoes, sailboats, and hammers remain commonly used even in the most advanced societies, despite the advent of much more spectacularly powerful technologies.

The situation is complicated by a frequently recurring pattern in which relatively recent technological forms disappear while much earlier and simpler forms, being further removed from the forms newly introduced and thus better able to compete with them, retain a stable even if reduced existence. Thus, in the 1980s, we are much more likely to find people engaged in the ancient activity of horseback riding than to find them driving automobiles that date from the 1920s; 1980-model cars have displaced 1920 models much more than they have displaced the horse or the bicycle. Similarly, with new advances in medicine, it is comparatively difficult to find in the United States reliance on "outmoded" medical procedures of half a century ago, although procedures very much older such as faith healing manage to survive.

Sometimes an effort is made to modernize older types of technology that are challenged by newer types. The German military strategist Friedrich von Bernhardi insisted, prior to the First World War, that "the army cavalry must . . . be equipped and conversant with wireless telegraphy, telephones, signalling apparatus, and flying machines."[22] And in 1940, with World War II already well under way although the United States was not yet directly involved, a spokesman reported on progress in modernizing the military use of horse cavalry in the United States armed forces:

> The illusion prevails that horse cavalry is outmoded. . . . Let us dispel these illusions once and for all. . . . While the European nations, hampered by . . . European traditions . . . were clinging to the mounted charge, the American Cavalry . . . has . . . led the world in adapting itself to the conditions of modern war by abandoning foolish mounted attacks and using speed of movement to put the fire power on the ground. . . . Whereas horse cavalry can flow over all types of terrain . . . mechanized cavalry is going to be chained largely to the roads. . . . Our cavalry is modernized and keeping pace with all developments. We are particularly fortunate in having great resources both in horses and motors. There are more than ten million horses in this country.[23]

An older technological form may sometimes even benefit temporarily by associating with a newer form that will ultimately displace it. Thus, steamers, which ultimately replaced sailing ships, initially helped them by taking them into and out of harbors and thus contributed to the solution of one of the difficult technical problems of the Age of Sail.[24]

In countries that remained behind during the Industrial Revolution and whose leaders are now seeking to catch up, we commonly find startling contrasts between older and newer technologies flourishing side by side in different sectors of the economy. China, for example, is a country in which bricks are often transported to construction sites by wheelbarrow, in which food is often transported into cities from the surrounding countryside in carts pulled by human muscles; but China has also produced her own nuclear weapons and launched her own satellites into Earth orbit. Sometimes very old and very new technologies are brilliantly integrated in these developing or "transitional" societies, as when television programs beamed to India from orbiting satellites are received in remote villages in which electric power is provided only temporarily by wheels turned by human muscles. Quite often in these societies we also find technologies that are intermediate between advanced and backward or that are modern but outmoded: for example, computers that are ten or fifteen years behind advanced standards.

In the United States and other advanced Western societies, there has appeared in recent years among some segments of the population a negative reaction against some advanced technology and a striving to develop instead what has come to be known as "alternative technology" or "appropriate technology." A major source of inspiration for the appropriate-technology movement has been the writing of E. F. Schumacher, who pointed out the need for methods and equipment that are cheap, that can be applied on a small scale, and that are "compatible with man's need for creativity."[25] For administrative purposes in the United States, the National Science Foundation has defined appropriate technologies as

> those which possess many of the following qualities: they are decentralized, require low capital investment, are amenable to management by their users, result in solutions that conserve natural resources, and are in harmony with the environment; they are small or intermediate in scale, take into account site-available natural and human resources, and are more labor than capital-intensive.[26]

One difficulty with this concept has been identified by Harvey Brooks, who pointed out that early in this century the automobile resembled appropriate technology in its interface with consumers (i.e., it was cheap and easy to repair) but possessed features diametrically opposed to appropriate technology as far as its production was concerned (i.e., auto production was centrally organized with assembly-line methods). Furthermore, Brooks pointed out that these two aspects of automobile technology were interdependent: cheapness depended on economies of scale in production; and the production methods thus involved were dependent, in turn, on a mass market.[27] The ill-fated attempt to have steel produced in millions of backyard furnaces in China during the Great Leap Forward of 1957–61 illustrates a reversal of this situation: it involved decentralized production, but presumably the steel thus produced would have to be collected and used in a much more centralized manner.

We may also note that the moral evaluation that has come to be linked with the concept of appropriate technology may not always be relevant. If nuclear weapons become cheap enough and easy enough to produce, so that small groups and individuals can acquire their own and thus eliminate what has hitherto been a monopoly of these weapons by central governments, we would have a phenomenon characterized by several of the basic features of appropriate technology. Similarly, if instead of, or in addition to, having wealth redistributed from rich to poor through centralized government programs, we were to develop a kind of do-it-yourself redistribution system involving large number of robberies by poor people stealing from the rich, we would also have some important features of appropriate technology manifested. These examples are cited here as a reminder that decentralization, small-scale operations, and individual initiative, in contrast to centralized decision making and centralized operations, do not always constitute "appropriate" technology.

Appropriate technology may be utilized in "backward" situations in which advanced technology is not yet established or where advanced technology already flourishes. In the latter case, it may be intended as a supplement to advanced technology or as a replacement for it. The total or near-total replacement of our familiar modern technology by appropriate technology would entail a fundamental collapse of our present economic and political systems. Presumably this will not happen, unless our economic and political systems collapse from other causes, thus giving appropriate technology a renewed opportunity to fill the gap thereby created. Barring such a disruptive transformation, appropriate technology may be useful in certain of its forms, filling in gaps in our existing technological system and replacing highly centralized technologies in particular contexts in which decentralization has special advantages.

The future development of technology, like the future of society as a whole, cannot be predicted in detail with reasonable confidence. Technological feats far transcending any yet achieved have now become plausible possibilities (even though not certainties by any means) for the next few decades. We *may* see exciting new space explorations, contacts with intelligent life on other worlds, important new conquests of disease, a slowing down of the ageing process, new mechanisms of human reproduction (e.g., cloning, in which a person would be created with only a single biological parent), and radical new types of transportation and communication. Of course, progress in these directions could be disrupted by social resistance, war, political disorder, or environmental crisis. Thorstein Veblen suspected that modern technology might be intrinsically unstable: that in the near future its component parts may "become so closely interdependent and so delicately balanced that even the ordinary modicum of sabotage involved in the conduct of business as usual will bring the whole to a fatal collapse."[28] However, despite the seemingly fragile and vulnerable character of our modern technological system, which requires precise coordination among numerous components, this system has actually demonstrated remarkable recuperative powers. For example, the major belligerents of World War II recovered and surpassed their prewar populations and levels of production within a very few years, in striking contrast to the exceedingly slow recovery following termination of the Thirty Years War three centuries earlier in 1648. To a considerable extent, the recovery after World War II can be attributed to assistance pouring into devastated areas from parts of the world that remained largely untouched and affluent, notably the United States. However, another quite different factor may have also been involved: given a population with modern skills and accustomed to modern technology, intense destruction—provided it is not so intense as to threaten the continuity of the society itself—may stimulate technological advancement by eliminating outmoded facilities and procedures.[29]

5

Variations in Technological Style: The Case of China

In addition to recognizing certain general *types* of technology, which previous chapters have discussed, we may recognize also within each major country a distinctive national *style* with respect to technological development and organization. Like other cultural phenomena, such styles tend to be taken for granted and thus remain unnoticed by those who participate in them, except to the extent that observation of contrasting styles abroad or of contrasting tendencies at home provides a basis for comparison.

The technological style prevailing in the People's Republic of China will be described here in order to provide a basis for a clearer understanding of the quite different styles prevailing in the more advanced countries of the West. This description will be based partly on the author's observations in China in February 1979.[1]

China does not provide the greatest possible contrast with the United States or other advanced Western countries. For that we should examine a small primitive society instead. China and the United States share important common features: they are both large societies with respect to territory and population, China being the most populous country in the world; they are both nuclear powers with satellite launching capacities, the United States being notably advanced in these respects; they are both ideological leaders on the world scene, with China as the center of a Communist movement rivaling that of the Soviet Union, and the United States as the chief exemplar of Western-style constitutional democracy with limited and decentralized government; both share important cultural (including technological) features that are characteristic of the contemporary world.

This background of shared features permits crucial differences to emerge clearly. China is a poor country, and the United States is a rich one; China is under single-party Communist rule whereas the United States is a strongly anti-Communist multi-party constitutional democracy; China has a centrally directed economy whereas the United States has an economic system emphasizing competition among giant privately owned corporations; China has never absorbed the individualistic ethos that the United States acquired from its European cultural background and from the experience of resettlement on a new continent; and the cultural traditions of the two societies are highly divergent in numerous other ways as well.[2]

China had a highly developed ancient civilization, impressive enough to elicit the amazed admiration of the Italian Marco Polo, who visited there between A.D. 1275 and 1292, and who returned to Italy with tales of such wonders as asbestos, paper currency, and the burning of "black rocks" as fuel.[3] About that time China had perhaps the most advanced technology in the world. Numerous technological innovations are reported to have originated in China, including gunpowder, paper, the mariner's compass, porcelain, silk, cable suspension bridges, movable-type printing, and canals with lock gates.[4] However, for reasons that are not entirely clear and that remain a subject of scholarly debate, Chinese technology ultimately failed to continue its advance.

China did not participate in the seventeenth-century Scientific Revolution or the subsequent Industrial Revolution; these developments, which ushered in the age of *modern* science and technology, appeared only in Western (European) civilization. For some time the Chinese leaders remained unaware of what was happening and of its implications. In 1793 the Chinese emperor, Ch'ien Lung, apparently thought that the free samples of British products sent to him by King George III were intended as tribute, and replied to the king's suggestion for trade between Britain and China by announcing that "we possess all things . . . and have no use for your country's manufactures."[5] Beginning with the Opium War in 1839, China was severely encroached upon by foreign powers, and a few Chinese came to recognize that their country would need Western technology to defend itself against the West. The slogan "Chinese learning as the basic structure, Western learning for practical use" represented one ultimately unsuccessful late–nineteenth-century attempt to resolve the ensuing dilemma. Chinese national unity and independence were substantially restored only with the immense social upheaval associated with the founding of the People's Republic of China in 1949 under the leadership of the Communist Party and its chairman, Mao Zedong. The one feature of Chinese material technology that a visitor from the West is most likely to notice is the overwhelmingly heavy reliance on human muscles: one sees carts loaded with food and other merchandise being pushed and pulled through city streets; people carrying loads attached to poles balanced over their shoulders; bricks being carried to construction sites by wheelbarrow; and rush-hour bicycle traffic in major cities. I recall, in particular, men riding tricycles pulling carts loaded with vegetables, streaming into Shanghai after dark from neighboring farms, and two men walking through a street in a south Chinese city, carrying a woman on a stretcher.

Of course there is nothing uniquely or distinctively Chinese about bicycle riding or the pulling of carts loaded with merchandise or the use of wheelbarrows (even though the wheelbarrow was reportedly invented in ancient China). These human-muscle technologies are commonly employed in many parts of the world (including the most technologically advanced parts, although they are not *dominant* there). In countries with ancient civilizations, which fell behind technologically and whose leaders now seek to catch up with the most advanced countries—and this includes China and India as well as a number of other Third World countries—such human-muscle technologies that lack fundamental *local* significance by virtue of their "universal" character tend to coexist with material technologies of several other types. First, there is technology which reflects distinctive local traditional cultures; for example, traditional Chinese medicine, which still flourishes with encouragement from the new Communist regime.[6] Second, there is modern technology, imported from the West or developed locally from Western models. Within this latter category, there is an especially wide range of variation. It includes technology that, although modern, is several decades old (or older) and was introduced in the days of colonialism by Westerners, largely in connection with efforts to control and exploit the population, for example transportation systems and facilities originally built primarily to satisfy Westerners' needs. It also includes at the opposite extreme material symbols and protectors of China's newly reacquired national independence, such as nuclear weapons and satellite-launching facilities (homemade by Western-trained Chinese experts) and jet liners imported into China by the new government but flown by Chinese pilots and serviced by Chinese mechanics.

These differing technological phenomena have sometimes become integrated in diverse ways. Integration of traditional Chinese medicine and modern medicine is illustrated by the use of ancient acupuncture techniques for modern purposes, such as anesthesia during modern surgical operations; acupuncture is perhaps three thousand years old but was first used for anesthesia during surgery in 1956.[7] Integration of simple manual labor and modern technology is widespread: "Modern industrial processes may make use of supplies brought in by hand-drawn carts, and automated production methods may employ materials that have been prepared by extensive hand labor."[8]

However, integration among different kinds of technology in China is highly imperfect. Although traditional Chinese medicine and modern Western medicine have been integrated in certain respects as indicated earlier, they nevertheless compete with each other. In some hospitals, patients can choose between them, although apparently each alternative is "impure," containing at least a bit of the other, and traditional and modern physicians do not always get along well together. The overwhelming reliance on ordinary human muscles in a country that has produced its own nuclear weapons and put its own satellites into orbit reflects a huge internal disparity, which is subject to the same conflicting interpretations as the technological disparities among nations. Perhaps, on the one hand, the technologically advanced sectors of the Chinese economy represent centers from which advanced technology will spread into other sectors. Or perhaps advanced technology in some sectors may help to perpetuate backwardness in others; for example, the diversion of resources into development of nuclear and satellite-launching capacities for military and international-political purposes may have interfered with efforts to elevate other aspects of the economy above the level of reliance on human muscles.

Chairman Mao's approach concerning the importation of Western technology and culture was formulated by Mao himself as follows:

> China should assimilate from foreign progressive cultures in large quantities what she needs. . . . However, we can benefit only if we treat these foreign materials as we do our food, which should be . . . separated into nutriment to be absorbed and waste matter to be discarded; we should never swallow anything whole or absorb it uncritically . . . [9]

This statement represents an ideal that was not always achieved: in various situations foreign technology was either blindly resisted or uncritically absorbed. There has also been a post-Mao policy shift (the duration of which cannot be predicted as these words are written) in favor of large-scale importation of technology, even if some foreign values that the regime does not like are thereby allowed to seep in.

Selectivity and variation in degree of selectivity have also characterized Maoist and post-Mao policies toward China's traditional technology; pride in China's ancient achievements has been mingled with the revolutionary urge to repudiate humiliating aspects of the past. Some components of traditional Chinese medicine have paradoxically retained respect from Communist revolutionaries, apparently because (1) traditional medicine is sometimes effective (even if, as skeptical outsiders insist, its effectiveness is at least partly psychogenic); (2) it is distinctively Chinese and thus represents an area in which China may claim to lead the rest of the world; (3) it is perceived as a product of trial-and-error among China's peasants over centuries of time and thus has a social origin that a peasant-based Communist regime would be inclined to look upon favorably (although *some* aspects of traditional Chinese medicine reflected the interests and values of the old upper class and are thus viewed less favorably by today's Communists); and (4) it contains features that, although originally interpreted within the framework of traditional Chinese assumptions and values, are also interpretable in terms of Marxist-Maoist dialectical concepts; for example, the antipain effects of needle acupuncture have been explained as representing a reinforcement of the body's natural dialectical negation of pain.

The modern technology that does exist in China is quite limited and largely outmoded by advanced world standards, with a few notable exceptions. China's nuclear weapons and means for delivering them are not at all impressive when compared to American and Soviet achievements in this field, although they might nevertheless have the intended deterrent effect on foreign powers. Although the Civil Aviation Administration of China flies modern jets imported from the United States, Britain, and the Soviet Union as well as some older planes, China's airports appear to be nearly empty much of the time. When I arrived at Beijing (Peking) airport in 1979, a Romanian jet was the only other plane visible at the entire airport. Although there are enough automobiles to produce very minor traffic jams in large cities at rush hour, the number of autos is miniscule by American and western European standards. These are all publicly owned by some unit of government or some collective enterprise; there are no privately owned cars in China, although some privileged persons have cars at their disposal for official—and sometimes, illicitly, for personal— business. Although the best modern medical care available in China is very good, much medical care there is of poor quality. Despite the "barefoot doctors" who are selected from among the peasants and who return to the countryside after brief medical training, many millions of peasants have extremely limited medical care available to them.

From a long-range point of view, the reliance on human muscles as a central source of power in China is an indication of "backwardness." Chinese Communist leaders have openly described it as such, blaming it on imperialism and feudalism under earlier regimes; and tour guides are sometimes embarrassed when visitors comment on it. However, in terms of short-run considerations of efficiency, this phenomenon appears to be a rational utilization of China's surplus of low-skilled manual labor in a way that compensates for acute shortages of other resources.

While walking shortly after dawn in a Chinese city in 1979, I happened to see a garbage collector going down the street with her wheelbarrow, broom, and dustpan. As she moved down the street, residents came out of their modest homes and dumped their garbage unwrapped in the street. She scooped it up; and when the wheelbarrow was full, pushed it back in the direction she had come from, returning later with an empty wheelbarrow to finish her work. In the United States, this garbage-collection method would be inappropriate not only because labor is expensive and because garbage bags and pails are cheap, but also because garbage lying in the streets unwrapped would interfere with traffic; would be a health hazard and an eyesore; and would attract—and be dispersed by—roving bands of dogs. In the Chinese situation that I observed, arrangements for the collection of unwrapped garbage by wheelbarrow appeared, by contrast, to be eminently reasonable: labor was cheap; materials from which garbage pails and bags would be made were expensive; dogs are not permitted and are generally not found in Chinese cities; the frugality associated with poverty drastically limits the quantity of garbage and prevents the inclusion in it of edible materials; the garbage was placed in the street only when the garbage collector was in sight. We have here an illustration of the possibility that extremely simple technologies that would be out of place in most contexts in advanced societies may be entirely appropriate under some very different conditions, and the possibility that simple technologies may be integrated with other social arrangements in comparatively stable patterns.

On the other hand, Chinese technology in some important aspects is "backward" even when evaluated in terms of criteria specifically appropriate to the Chinese context. Arable land is limited in China, whereas agricultural labor is generally plentiful except at certain critical times such as the harvest. It is thus reasonable from a short-run point of view that Chinese agriculture, in which the overwhelming majority of the population is engaged, should be evaluated in terms of productivity per hectare or per acre rather than in terms of productivity per worker. The latter is a more suitable criterion in North America, where land is comparatively plentiful and labor is expensive. But Chinese agriculture has comparatively low productivity even when measured according to the per-hectare criterion that is most appropriate to China's circumstances. And from a long-range standpoint, one must see the shortage of arable land not merely as a condition to which adjustment must be made, but also as largely correctible consequence of human abuse of the land in the past. Similarly, the surplus of low-skilled labor is a condition that might possibly be transformed in the future through birth-control measures and through education, both of which are strongly emphasized by the regime.

The technological backwardness of China is reflected in living conditions that are extremely Spartan by American or western European standards: even model apartments that foreign visitors are shown commonly have no stoves, refrigerators, running water, toilets, carpets, or telephones; have extremely limited lighting that is not adequate for reading at night; and have very limited heating even where winters are cold. For residents in major cities, adequate food is available most of the time, although variety is limited. Many urban residents have been able to acquire wristwatches, transistor radios, bicycles, and even television sets. Conditions in the countryside where most Chinese people live are generally much more difficult. The effects of poverty are mitigated for some people to some extent by access to relevant public or shared facilities; for example, telephones in residential streets, with arrangements for calling people to the nearest street phone from their homes; and toilets, running water, stoves, and television sets that may be available on a shared basis to people who do not have these facilities within their own private residential units.

Poverty has meant that certain forms of waste that are common in the affluent society of the United States are generally not found in China. One does not find in Chinese garbage the vast quantities of food and still-usable goods of other sorts that Americans commonly throw away. Chickens but not dogs roam through Chinese city streets and courtyards; they find food for themselves by foraging, and they constitute food for humans, thus helping to recycle bits of edible garbage that might otherwise be wasted. However, such indications of frugality are only part of a larger and more complex picture. There has been not frugality, but rather extravagance, in the use of human muscle power, which is understandable and reasonable in the sense that this is one resource of which China has an excess supply. But scarcity also tends, paradoxically, to produce a special kind of extravagence, as illustrated by the classic story of the farmer who can only save himself and his family from starvation by eating the seeds that he "should" be planting to produce next year's crops. In China, as in other poor countries, urgent situations have often forced people to neglect certain long-range considerations for the sake of immediate remedies, even though the regime—like Communist regimes elsewhere in the world—is fundamentally committed to the idea that contemporary satisfactions should be sacrificed as much as possible in pursuit of long-range objectives. Extravagance produced by scarcity rather than by abundance is illustrated by the widespread destruction of trees to obtain firewood, with disastrous environmental consequences—a practice that the regime is energetically counteracting with a major reforestation effort. But the most interesting and most tragic form of extravagance has involved the waste of human resources, especially of scarce talent. During the years of Chairman Mao's power (1949–76), and especially during the Cultural Revolution in the late 1960s, there was conspicuous waste of extremely scarce human resources, with highly educated people possessing skills that are rare in China (including scientists and other intellectuals) being sent out to do ordinary manual labor among the peasants, with their special skills remaining unutilized and rusting away from disuse.

Deliberate waste of a scarce and badly needed resource such as highly skilled personnel might appear to be irrational; and this has, in fact, been condemned in a major reversal of policy in China since Mao's death in 1976. However, it is not difficult to understand at least some of the reasons why such waste was encouraged under Mao's leadership. First, although scarcity of a valued resource may generally increase its value, another consideration also applies when this scarce resource consists of people of a certain kind: relatively small numbers will mean not only *increased* bargaining power as a consequence of scarcity, but also *reduced* "voting strength" or its equivalent; Chinese intellectuals were thus weakened in one respect, even while potentially being strengthened in another respect, by their status as a small minority. Second, China's need for intellectuals was minimized in Mao's time, with the assumption that ordinary workers and peasants without specialized scientific or technical training could replace them in their functions: while scientists were doing manual labor, peasants were encouraged to do "scientific research." This interpretation was reinforced by a change in the kind of scientific and intellectual activity that was valued, with great emphasis being placed on immediate practical utility, which is a meaningful criterion to peasants, and with emphasis removed from abstract theory, which is of interest only to intellectuals and not to peasants.

In 1979 one could mail a local letter in Beijing in the morning and expect it to be delivered that same afternoon. Tour guides with whom I spoke assumed that in the United States mail delivery would be even more rapid. The superiority of local mail delivery in this respect in a "backward" country as compared with an "advanced" country is actually easy to explain: (1) Mail delivery is more important in China than in the United States, because China lacks a highly developed telephone system. (2) Cheap labor in China makes rapid local mail delivery more feasible there than it would be here, at least in crucial locations. We should be prepared to find, more generally, that backward countries may sometimes outperform more advanced countries in particular areas of technology that are especially important to them and in which their backwardness may provide paradoxical advantages (e.g., a cheap labor supply). Extreme backwardness may also provide a degree of flexibility that permits a backward society to bypass other societies that are more advanced but that have acquired vested interests in established systems. Thus, it was easier to introduce electric street lighting into cities that had not yet lit their streets by nonelectric means, and a system of simplified writing would be easier to introduce into a society in which writing is a novelty than into one in which an older writing system is already deeply entrenched. On the other hand, backwardness can also become involved in a self-reinforcing vicious circle.

The surplus of low-skilled labor in China has encouraged a disinterest in laborsaving devices that have been strongly emphasized in the most advanced Western economies. This condition, combined with the Maoist striving for mass participation and monolithic unity, encouraged—during Mao's dominance—projects in which huge numbers of people, inspired through major propaganda efforts and carefully coordinated, accomplished tasks that would have been accomplished in the West with a much smaller number of people and with laborsaving materials. The previously mentioned 1957 "war against the sparrow," in which the entire population of Beijing was mobilized to prevent sparrows from taking refuge anywhere and to keep them flying until they collapsed from exhaustion, provides a clear example.

Novel methods for organizing mass activities have constituted perhaps the most important of all recent Chinese technological innovations. Often, mass activities in China have focused on tasks (e.g., killing sparrows, building dams) that in other countries would have been performed in other ways, with a small number of people armed with relevant equipment that the Chinese masses lacked and for which they substituted force of numbers. In at least one case, however, mass activities were centrally involved in a unique and distinctive achievement. On February 4, 1975, Chinese seismologists made the world's first successful prediction of a major earthquake. Chinese quake-prediction procedures involve the mobilization of many thousands of people on a part-time basis to monitor numerous and widely scattered observation stations and to look for signs of prequake animal behavior (unusual excitement) and for sudden flooding or drying of wells (these conditions have been found to be indicators of forthcoming quakes).[10] According to one estimate, the quake-predicting effort in the 1970s came to involve about ten thousand geological workers, five thousand observation points, and one hundred thousand peasants and other lay observers.[11] Quake prediction is one of the few research areas in which Chinese investigators in recent years advanced clearly, even if only temporarily, beyond their colleagues in other countries (another such area involved the synthesis of insulin, first achieved in Shanghai in 1965).[12] Quake prediction happened to be an area in which China had a traditional interest (the world's first seismometer was built there in ancient times) *and* in which modern China's emphasis on mass activity was relevant in a way that gave China's quake-prediction investigators an advantage over their colleagues elsewhere. It was also an area in which the Chinese leadership had a special interest in encouraging mass participation. Traditionally, great natural disasters have been interpreted in China as indications that Heaven was displeased with the existing regime. Thus after a solar eclipse in BC 178, Emperor Han Wen-ti admitted that "Heaven has reproached me."[13] The Chinese communists have an obvious interest in encouraging popular understanding of more modern explanations of earthquakes.

By Western standards the Chinese people appear to be extremely tightly controlled: unable to move freely from one part of the country to another, to change jobs without permission, or to make fundamental criticisms of the regime (except on special occasions when the restraints on free speech have been temporarily lifted, giving hidden opponents of the regime a temptation to expose themselves). Furthermore, controls are extended into spheres of life that Westerners have come to regard as peculiarly private, as illustrated by meetings held in workplaces in which an allocation of pregnancies is agreed upon among the workers' families. There has also been a notably widespread use of certain forms of material technology related to propaganda and control, for example, an estimated 140,000,000 loudspeakers in rural areas as part of the "national wired loudspeaker system."[14]

However, controls designed primarily for health and safety purposes appear to be *less* prominent in China than in the United States. I noticed that on three plane flights in China, no airline employee asked us to fasten our seat belts or to follow any other regulations or procedures (and some Chinese planes do not even have seat belts.) Similarly, factory managers are not subject to regulatory controls as complex as those imposed on their American counterparts in relation to worker safety and pollution control; and as of 1979 cigarette smoking was openly encouraged in China, apparently without even the minimal inhibitions that have been gradually building up in the United States. There is a *double* paradox here: laxity in health-and-safety controls coexists both with extremely tight controls with respect to various other aspects of life *and* with major efforts to improve health in other ways (e.g., through the system of barefoot doctors).

My own somewhat speculative interpretation of this situation is as follows:

1. Chinese controls tend to be of a type fundamentally different from American controls: there is emphasis on "concrete" controls organized around major goals (increasing productivity, strengthening national defense, protecting the regime against internal opposition, etc.) rather than on "abstract" controls involving complex written regualtions of which the underlying purpose may be obscured.

2. A few uncontrolled aspects of life may help to sustain morale and thus make heavy controls in other areas more stable and effective.

3. Certain kinds of control are apparently regarded as too costly in China, even though they may be considered cost effective in more advanced countries.

4. Expenditures on health measures are more likely to be undertaken in China if clear political objectives can be achieved at the same time. The barefoot-doctor program received support not only for medical reasons but also because the barefoot doctors are to some extent political and ideological agents of the regime in the countryside, as well as medical agents. Protecting the ears of workers in noisy factories does not have similarly clear and direct political implications.

China's national technological style has been shaped by (1) basic economic conditions including widespread poverty, a scarcity of many resources crucial to advanced material technology, shortages of personnel with modern skills, and an abundance of cheap and low-skilled labor; (2) an ancient cultural tradition emphasizing Chinese superiority, collectivistic rather than individualistic values, concrete rather than abstract thought, a sharp prestige differential between mental and manual labor, and minimization of the division of labor in most other respects; (3) post-1949 commitments to the Maoist version of Communist doctrine emphasizing dominance by "the masses," monolithic unity, centralized controls over individual behavior, economic egalitarianism, subordination of abstract expertise to practical experience, and subordination of immediate benefits to long-range goals; and (4) pragmatic concerns with matters such as national defense, economic development, and alleviation of immediate economic distress, which have recently come to take precedence over concern with doctrinal matters.

These diverse considerations, reinforcing each other in some instances and having mutually incompatible implications in others, interact to give China its distinctive technological style, which contrasts strikingly with the technological styles of other countries with which readers of this book are more likely to be familiar and which thus provides a basis for understanding the latter more clearly.

6
Technological Innovation

The Concept of Technological Innovation

Some changes in society take place accidentally, without being planned. Thus, where a free-market economy prevails, changes in price levels may occur through the impersonal functioning of the market as a consequence of the relationship between supply and demand, which no single individual or group is sufficiently powerful to control. Similarly, long-range changes in language—in grammar, vocabulary, spelling, pronunciation, etc.—are continually taking place without being planned. In some cases these unplanned, unintended, accidental changes are remarkably orderly and systematic. For example, in the Germanic languages, over a period of centuries every "p" was changed to "f" and every "t" to "th".

We are to focus here *not* on changes of these sorts but rather on deliberate, planned changes, and not on all planned changes but only on those that represent new means for achieving clearly identifiable ends and that are thus classifiable as *technological innovations*. The boundary between technological innovations in this sense and unplanned changes is not at all sharp and clear. At least two sorts of conceptual complexities prevent us from drawing this boundary very sharply:

1. Some items of new knowledge that ultimately come to be applied in a fully deliberate way to the solution of practical problems have their origins in purely accidental observations. For example, the accidental darkening of some "unexposed" photographic film in the late nineteenth century led at least one purchaser to complain to his

supplier about defective materials; but another purchaser, Wilhelm Roentgen, explored the problem more deeply and found that the plates were being darkened by a hitherto unknown form of radiation, which became known as "Roentgen" rays and later as "X rays."[1] Similarly, the accidental observation that the presence of certain molds killed certain bacteria led Alexander Fleming to trace the identity of the substance in the mold that was responsible for this result and hence to the discovery of penicillin.

2. In some cases it is difficult to tell in looking back on an episode in which a desired result was achieved the extent to which the episode was planned in advance. Thucydides reports that about the year 423 B.C., the Spartans offered freedom to any of their "helots" (serfs) who could show that he had performed especially meritorious service. However, this was merely a device for smoking out potential agitators within the helot population: about two thousand of them who stepped forward to claim their freedom disappeared and were apparently executed.[2] About 25 centuries later, in A.D. 1957, the Chinese Communists under Chairman Mao invited criticism of their leadership under the slogan "Let a hundred flowers bloom; let a hundred schools of thought contend"; but shortly thereafter talk about a hundred flowers gave way to angry denunciation of "poisonous weeds." Perhaps this episode, like that which Thucydides describes, represented a deliberate plan to smoke out and trap potential rebels; but there is also a very different possible interpretation: the invitation for a hundred schools of thought to contend may have been offered in the genuine expectation that the emerging debate would remain within narrow limits acceptable to the Communist leaders; and when the debate actually moved far beyond such limits, the leaders' attack on "poisonous weeds" may have been merely a response to an unexpected situation rather than part of a preplanned strategy.

The concept of technological innovation involves not only a problematic distinction between innovations that are planned and those that are not, but also an equally problematic distinction concerning the extent to which various phenomena are truly innovative. Would-be innovators sometimes find that the products of their researches are not as new as they had hoped and that their discoveries or inventions have already been produced by others. But even when a genuine innovation has taken place, an inquiry into its historical background will often show that the innovative aspects are less prominent than they appear superficially to be. Thus, the machine gun, which made possible the firing of many bullets in rapid succession without reloading, appeared to society generally as something basically new; but from a broader historical perspective it was merely an application to guns of a concept dating back to the ancient *polybolos*, which fired arrows in succession.[3] We should not be led by dramatic *effects* of innovations to conclude that the innovations themselves must be fundamental; sometimes even a minutely small innovation can have major effects.

In contrast to the widespread tendency to exaggerate the newness of innovations, we sometimes find important innovations produced by people who think erroneously that they are copying others and who remain largely unaware of the extent to which they are really pioneering new ground. Two illustrations of this phenomenon are particularly interesting:

1. Our first illustration pertains to the achievement of the Cherokee Indian Sequoyah, already mentioned in Chapter 2. Sequoyah *thought* he was imitating the white man when he constructed a system of writing for his people. However, he did not understand enough about the white man's alphabet to know that he was not merely imitating it but was actually improving upon it: his script apparently made it possible for Cherokee youth to learn to read simple messages in a very few days, something that white youth using their alphabet could not usually do.[4]

2. In the late nineteenth and early twentieth centuries, young Americans who wished to have the best possible education in diverse fields of science went to study in Germany, which was the world's leading center of science at the time. Some of these scholars returning to the United States sought to build in their own country, universities organized fundamentally according to the German model. However, their status as visitors or guests in German universities rather than as full-fledged participants, apparently gave them a limited view of the German system. This, combined with certain cultural differences between Germany and the United States, led them to reproduce the German system "incorrectly" and to improve upon what they were imitating, without full awareness that they were doing so. In German universities of the time, there were no academic departments. Each discipline represented in the university had a "chair" (i.e., a permanent professorial position occupied by a single incumbent on a permanent basis). The single professor occupying a chair had various subordinates who actually did much of his work, but these were employed in the professor's own institute and were not full-fledged members of the university faculty. American scholars returning home from Germany

imitated this system by establishing academic departments of the sort familiar to us today.[5]

Aspects of Technological Innovation

Technological innovation as here defined has several aspects or phases, which have been delineated and described in diverse ways. These may be briefly identified as follows:

1. Discovery and invention are commonly described as distinct processes. We would ordinarily say that Newton discovered his law of gravitation and that Bell invented the telephone. However, the distinction is not a clear one. Any invention may be regarded as a discovery of a means by which some given end may be attained: thus, Bell discovered a means for making telephones even if we ordinarily say that he invented (rather than that he discovered) the telephone itself. Correspondingly, any discovery may be regarded as an invention of an idea that leads to improvement in our knowledge: thus Newton invented the idea of the law of gravitation even if we ordinarily say that he discovered (rather than that he invented) the facts concerning gravitation itself.

We may reasonably use either of these terms, "discovery" or "invention," depending on which aspect of the given situation we seek to emphasize. When the most relevant aspect is the finding of new information, especially about the way in which the natural world functions, "discovery" may best be used. When attention shifts to focus on new ideas for ways to solve practical problems (other than the problem of understanding the natural world), "invention" may be more appropriate.

Given this contextual distinction, discoveries in a given field will appear to be prerequisites for inventions in the same field: first came the discovery of the existence and the potential of nuclear energy and only later came the invention of the atomic bomb—and this sequence presumably could not have been reversed. This conception of discovery as having a logical priority over invention is reasonable provided certain qualifications are allowed for:

—There is generally not a one-to-one correspondence between discoveries and subsequent inventions based upon them: one discovery may suggest many diverse practical applications (e.g., the innumerable uses of electricity) and a single invention (e.g., the automobile) may be based upon a large number of preceding discoveries.

—A discovery does not always necessarily lead to subsequent inventions based upon it; the possible existence of a discovery that remains "useless" on a more or less permanent basis cannot be ruled out.

—Some practical inventions, especially ancient and comparatively simple ones (e.g., the bow and arrow) do not appear to be based directly on explicitly identifiable preceding discoveries, even though they are based on intuitive understandings of the way particular aspects of the world actually function.

—A practical invention does not necessarily require knowledge concerning all the mechanisms by which the invention actually works, or knowledge of all causal elements in the situation that the invention is intended to change. Thus, one might invent a method for curing a disease even while the cause of that disease and the way in which the cure produces its results remain largely unknown. In fact, cases of this sort are commonplace: aspirin was used for years without anyone having a clear conception of the mechanisms by which its effects were produced, and for many centuries people knew that metals could be hardened by pounding without having any idea as to why this is true (we know today that the pounding tends to rearrange the molecules).

—The practical interests of inventors and of persons who will utilize their inventions may sometimes play a part in providing the necessary motivation for inquiries that lead to scientific discoveries. However, the extent to which this happens and the extent to which it influences the course of scientific development have been controversial. An extreme point of view in this respect was offered by the Soviet scholar Boris Hessen, who later became a victim of one of Stalin's purges. In a paper presented at a conference in London in 1931, Hessen suggested that the flourishing of science in seventeenth-century England was largely inspired by the practical needs of a rising merchant class, which sought, for example, satisfactory ways to determine longitude at sea, to ventilate mines, and to predict trajectories of hurled objects.[6] Hessen believed, for example, that Newton's law of gravitation was partly a response to the need of the new merchant class to be able to predict where a missile fired from a cannon aimed at various angles would land.

—Although practical inventions are dependent on previous discoveries (or on implicit understandings of the way relevant aspects of the world function), a discovery in turn may be dependent on previous inventions in fundamentally different areas of inquiry. For example, the discovery of microorganisms and the discovery of moons of Jupiter both required the prior invention of magnifying lenses, applied in microscopes in one case and telescopes in the other. Furthermore, many discoveries pertaining to early human history have been made possible by the previous invention of the method of "carbon 14 dating," based on developments in chemistry. (Carbon 14 dating is a method by which bones and artifacts from ancient times can often be dated much more precisely than was previously possible.)

2. A discovery may or may not be *accepted,* and an invention may or may not be *utilized.* These distinctions, however, are complicated not only by differing responses within relevant populations (some people may accept a discovery or utilize an invention whereas others do not), but also by a diversity of forms that acceptance and utilization may assume. Thus, a scientific proposition that emerges in a discovery process might sometimes be accepted *as* a useful basis for making predictions, even by people who are unwilling to accept it as actually "true." And even *non*utilization of some inventions may sometimes constitute a form of utilization: a country that does not explode nuclear weapons on enemy targets during wartime may thereby be utilizing such weapons for purposes of *deterrence.*

3. A discovery or invention arising in one group, community, or society may or may not be *diffused* or spread to contexts other than the one in which it first appeared. However, unless historical evidence is available, we might not be able to determine whether a technological item that appears in two different societies or social contexts has been invented independently in each of them or whether it has diffused to one of them from the other (or to both of them from a different common source.)

The concepts of discovery, invention, acceptance, and diffusion as employed in the study of social change were made familiar to social scientists through the writings of the anthropologist Ralph Linton.[7] In the administration of scientific and technological programs, a set of concepts is most often employed that differs somewhat from Linton's, and that centers around distinctions among "basic research," "applied research," and "development." These terms have been defined by the National Science Foundation as follows:

Research is systematic intensive study directed toward fuller scientific knowledge or understanding of the subject studied. . . . In *basic research* the investigator is concerned primarily with gaining a fuller knowledge or understanding of the subject under study. In *applied* research the investigator is primarily interested in a practical use of the knowledge or understanding for the purpose of meeting a recognized need. *Development* is systematic use of the knowledge and understanding gained from research, directed toward the production of useful materials, devices, systems or methods, including design and development of prototypes and processes. It excludes quality control, routine product testing, and production.[8]

For industrial research performed by commercial corporations rather than by university scientists or nonprofit organizations, the National Science Foundation uses a slightly different criterion for distinguishing basic from applied research: basic researches "do not have specific commercial objectives, although they may be in fields of present or potential interest to the reporting company," whereas applied research has "specific commercial objectives with respect to either products or processes."[9]

Presumably "basic research" in the NSF definition corresponds to "discovery" in Linton's sequence, although it refers directly not to discovery itself but rather to the process through which discovery hopefully occurs (but one must allow for accidental discoveries that are not based on research and also for research that fails to yield desired discoveries.) The outcomes expected from applied research may be variously classified as either discoveries or inventions. "Development" is a stage in the process of invention. "Acceptance," "utilization," and "diffusion" have no counterparts in the NSF sequence, although presumably the detailed operations that development commonly entails would not ordinarily be undertaken in the absence of evidence that the invention that is being developed will be utilized.

Conditions of Technological Innovation

Many investigators have sought to determine the effects of various working conditions on creative innovation among scientists and other researchers, for example, the effects of close supervision versus autonomy, large versus small work groups, and different procedures for allocating salaries and other rewards. Perhaps the best-known study in this area, conducted by Donald C. Pelz and Frank M. Andrews, found that research productivity is maximized when research workers are confronted by major *challenges* while also possessing considerable *security*.[10]

Still other studies have analyzed the conditions under which societies, as distinct from individuals, tend to be creative in particular ways; for example, Joseph Ben-David's study of the conditions that caused scientific activity to be heavily concentrated in northern Italy in the early seventeenth century, in England from the late seventeenth to the late eighteenth century, then in France until around 1830, in Germany from then until around the time of the First World War, and most recently in the United States.[11] William McNeill, referring to societal-level innovations generally rather than specifically to those of a scientific character, suggested that such innovations are most likely when a major change in the prevailing way of life coincides with exposure to *rival* ideas from abroad; the changing way of life creates a readiness to innovate, while rivalry among imported ideas discourages the blind imitation of any of them.[12] The achievement of the Cherokee Indian Sequoyah, discussed earlier in this chapter, and in Chapter 2, indicates that societal-level innovation is facilitated when there is sufficient contact with foreign cultures to permit the borrowing of general concepts (in Sequoyah's case, the concept of writing) but *not* sufficient contact to permit close imitation of details.

Individual-level and societal-level studies of the sources of innovation have not been adequately integrated. Furthermore, societal-level studies, which are of greatest relevance here but which suffer from our inability to manipulate societal-level variables for experimental purposes, have yielded especially unclear results, especially insofar as scientific innovation is concerned. Conditions that cause scientific activity to center in particular societies at particular times have not yet been adequately formulated in general terms, although convincing "reasons" why science has centered in any given society at any given time can always be found retrospectively.[13] The significance of freedom for science in particular has been a matter of continuing uncertainty. The idea that freedom is essential for science has often been emphasized, and detrimental effects of authoritarian regimes on science have often been noted.[14] On the other hand, the possibility has also been noted that regimes that restrict freedom generally but restrict it less in science than in other activities may thereby encourage talent to flow toward scientific work—talent that in a generally free society would very likely flow largely toward other occupations instead.[15] Another area of disagreement, or at least of sharply diverging emphasis, with respect to effects of societal conditions on scientific innovation, is indicated by the two following statements, both of which appeared in spring 1981: (1) "It is obvious that some kind of global unity and cooperation is a necessary condition for the unfettered growth of science in the future."[16] (2) "Our disordered global political situation, which encourages competition among nations in science as in other respects, and which permits science to center its activities under any of a wide variety of political jurisdictions, may thus play a crucial part in sustaining the autonomy and the progress of science."[17]

Without attempting to resolve these specific issues here, I shall suggest a general way in which we might reasonably approach the task of identifying conditions conducive to, and detrimental to, technological innovations of diverse sorts. As a basis for this, I shall briefly review certain historical trends concerning the rapidity with which technological innovation occurred.

1. Primitive peoples had technologies that were very limited and very slow in developing by modern standards (although highly advanced and rapid in developing by comparison with lower-animal "technologies").

2. With the emergence of early (premodern) civilizations, the pace of technological progress picked up, although such progress was highly irregular, with remarkably rapid advances in particular technological fields in narrowly delimited times and places, but also with occasional retrogression and with long periods of "stability" (to use a term that makes this condition sound good) or "stagnation" (to use a term that makes it sound bad). For example, the rapid appearance of several new technologies (metallurgy, irrigation, sailing ships, etc.) in Mesopotamia about 3000 B.C. was followed by an end to major technological innovation. A few centuries later the rapid development of Egyptian pyramid-building technology came to a close and was followed by thousands of years with no major additional progress in this area. And Chinese civilization, after brilliant technological achievements unmatched elsewhere, similarly reached a plateau in its technological development and rested there.

3. In recent centuries, technological progress has become increasingly rapid. This is especially clear with respect to science, considered here as one type of technology. Derek J. de Solla Price has found that the number of scientific publications has increased exponentially since the seventeenth century.[18] This finding is consistent with Lord Kelvin's statement in his 1901 Presidential Address to the British Association for the Advancement of Science, that "scientific wealth tends to accumulate according to the law of compound interest";[19] A. Conan Doyle's remark in 1894 that "knowledge begets knowledge as money begets money";[20] and a similar observation by Friedrich Engels earlier in the nineteenth century, that since the sixteenth century the development of science has "gained in force in proportion to the square of the distance (in time) from its point of departure."[21]

4. Rapid exponential growth in anything pertaining to human affairs should not be expected to continue indefinitely. We might

currently be at or approaching or very recently past a crucial transition point between several centuries of exponential growth in science and science-based technology and a period of slower and/or more irregular growth.

5. In any case, throughout the modern period of rapid technological progress, there have remained numerous "pockets of resistance" to change in diverse technological fields and diverse social settings.

This complex situation may be summarized as follows: (1) rapid exponential growth in technology as a whole and the world as a whole in recent centuries, and occasionally for brief periods in particular technological fields and particular countries in the distant past; and (2) comparatively slow growth most of the time in most technological fields and in most countries prior to the modern era and in numerous pockets of resistance today and possibly in the world as a whole in the future.

In interpreting this historical record we have a choice between two basic assumptions: (1) We might assume that established technologies tend intrinsically to be stable, so that technological progress when it occurs requires explanation, whereas an unchanging situation does not. (2) Or we might assume that technology has an intrinsic tendency to grow, so that an exponential increase in number of innovations is a normal phenomenon not requiring explanation, whereas a condition of stability must be explained in terms of factors that inhibit the growth that would otherwise occur. In either case, with either of these contrasting assumptions, our explanation may be formulated in terms of the presence or absence of *opportunities* for technological innovation and/or in terms of *needs* for such innovation or for its absence.

Needs For Technological Stability and For Innovation.

The familiar idea that necessity is the mother of invention embodies an important truth but is also incomplete and potentially misleading in several ways: (1) Necessity may be the mother not only of invention buy also of *resistance to* invention. (2) People tend to regard as "necessities" those things that they already have or that they perceive as reasonably possible for themselves to have, and their perceptions of these matters are influenced by past inventions—which means that, in a sense, *invention* may sometimes be the mother of *necessity* rather than the other way around. (3) Many inventions have not been at all "necessary" by usual standards: a book entitled *Inventions Necessity Is Not The Mother Of* mentions many of these including, for example, parakeet diapers.[22] Thus, on this topic we have a situation far more complex than a popular saying would suggest.

Reasons cited for opposing various innovations in the past often seem bizarre by today's standards. Street lighting in Berlin was "a presumptuous thwarting of Providence, which has appointed darkness for the hours of the night."[23] Anesthesia during surgery would "rob God" of the pleasure of hearing cries of pain.[24] Use of stagecoaches in England would reduce the sale of beer and ale along the roadsides and thus reduce royal revenues.[25] Acceptance of new "planets" (the moons of Jupiter discovered by Galileo) would upset the correspondence between the "seven" planets and the seven days of the week.[26] The internal-combustion engine in automobiles was not feasible because "you cannot get people to sit over an explosion."[27]

Resistance to technological change, even for seemingly absurd stated reasons, should not surprise us. Any such change will entail costs for some people, even if there will also be benefits for these same people and/or for others. Furthermore, these costs may involve much more than a threat to traditional values; in fact, the importance of traditional values as a source of resistance to change appears to have been overrated. It has been said that "the more a social change threatens or appears to threaten the traditional values of the society, the greater the resistance to that change and the greater its attendant cost in social and personal disorganization."[28] However, truly immense social and personal disruptions can ensue from changes in social conventions and laws that are only peripherally connected with "traditional values." Americans do not generally believe that driving on the right side of the road is intrinsically better than driving on the left, and Britons do not generally believe that left-side driving is intrinsically better. These are essentially matters of practical convenience: there must be *some* rule of the road, and in each country different historical developments have produced different rules. But any serious attempt to change the rules in any country, so that Americans would drive on the left or (more likely) so that Britons would drive on the right, would entail truly immense dislocations and arouse formidable resistance even though what are usually considered to be "basic traditional values" are not directly at stake. In American society at the present time, "basic values"—including religious commitments, conceptions of patriotism and national loyalty, ideas about the family and its proper role in society—appear highly unstable and amenable to rapid fluctuation, whereas numerous strictly technical arrangements—including measurement of distance in inches, feet, and miles; division of the day into twenty-four hours; driving on the right side of the road; shaking hands with the right hand rather than the left; adherance to the grammatical patterns of the English language—seem much more firmly entrenched.

A technological innovation may produce a stable outcome by satisfying a need in a way that does not produce any new needs. But often, instead, a new technological item creates a new need that calls for still more technological innovation. A new military weapon may create a demand for a new counterweapon. New ways of preventing or curing one ailment may cause people to acquire other ailments instead and thus create pressures for finding ways to cope with the latter as well. At the turn of the century, physicians writing in highly reputable medical journals described heroin as a drug that would relieve morphine addiction;[29] more recently, heroin addiction has stimulated a further innovation—the use of methadone, which eliminates the craving for heroin.

Opportunities for Technological Innovation

Concepts pertaining to this topic may be introduced with an example. The French mathematician Henri Poincaré described in these words how he made one of his discoveries:

> For fifteen days I strove to prove that there could not be any functions like those I have since called Fuchsian functions. I . . . tried a great number of combinations and reached no results. One evening . . . I drank black coffee and could not sleep. Ideas rose in crowds; I felt them collide until pairs interlocked, so to speak, making a stable combination. By the next morning I had established the existence of a class of Fuchsian functions . . .[30]

This particular discovery involved (1) a background or base of previous items (ideas), (2) conditions that permitted and encouraged these items to interact with each other, and (3) a selective process by which acceptable ("stable") combinations were distinguished from others and preserved. Many, and perhaps all, discoveries and inventions will fall into this general pattern.

Some innovations appear, at least superficially, to involve fundamentally new items rather than merely new combinations of old ones. The *tank* was invented merely by recombining previously existing technological items, but it appears somewhat more difficult to describe the invention of the atomic bomb in the same way. However, it is conceivable that even the most fundamentally novel innovations can be described as merely combinations of older items, and that our failure to see precisely how this interpretation can be made in some instances is merely a reflection of our own inadequate methodological concepts.

In any case, technological innovation always involves preexisting items even if it may also sometimes involve totally new ones. For any given technological innovation, the combination of prerequisite innovations constitutes the *cultural base*. Thus, the cultural base of the automobile consists of all the discoveries and inventions that had to come first before the invention of the automobile was possible—and, in enumerating these, we should not forget to include some that appear so obvious to us that they are easily overlooked (e.g., the wheel, which was apparently never invented in the western-hemisphere civilizations).

However, the concept of cultural base is complicated by the possibility that a single innovation might arise through any of several routes, some of which may be feasible only in certain social contexts. Sunspots were discovered independently by three astronomical observers in three different European countries, all in A.D. 1610–12. This happened because telescopes had recently appeared and were being utilized for the first time in astronomical studies. However, although the telescope thus appears to be quite clearly part of the cultural base for the discovery of sunspots *in Europe,* sunspots had nevertheless already been discovered centuries earlier in China without benefit of telescopes. Perhaps the European idea that heavenly objects were perfect and unchanging inhibited the discovery of sunspots (which appear as continually changing blemishes on the sun's surface) enough so that their discovery in Europe required assistance from the telescope that was unnecessary in China.

To the extent that innovation involves new combinations of old items, potentialities for innovation will be limited by the number of relevant preexisting items available for recombination. If this is the only limiting factor and if old items are retained as new ones are added, then the number of innovations per unit of time will tend to grow exponentially or in accordance with the compound-interest law. Such a pattern of growth has actually been reported in certain contexts, as already noted.

Of course, new items do not always leave older ones intact:

introduction of a new item will sometimes entail abandonment of older ones; and, if the new items are more fruitful than the old ones, the net result may be conducive to further innovation. In fact, the Scientific Revolution of the seventeenth century involved a series of such crucial abandonments of outmoded ideas: Copernicus abandoned the concept of the earth as the stationary center of the universe; Kepler abandoned the assumption that heavenly bodies move in perfect circles; Galileo abandoned the idea that a moving object continues to move only so long as it is pushed or pulled.

Historical barriers interfering with recombinations of technological items are well knows. Major insights emerged when the geometric astronomy of Greece and the arithmetic astronomy of Babylonia came together after the conquests of Alexander the Great in the fourth century B.C.;[31] earlier, such insights had been blocked by the cultural separation between these two astronomical traditions. The rise of modern science was apparently long inhibited by social-class barriers to communication between intellectuals and artisans, which made it difficult to combine the intellectual's abstract ideas with the practical, empirical knowledge that artisans accumulated.[32] In modern societies military and commercial requirements of secrecy often prevent discoveries and inventions from being fully communicated to others who might be able to build upon them.[33] As of several years ago, Chinese science was organized in a way that tended to fragment the scientific community, so that researchers in one institute would have relatively great difficulty in finding out about discoveries relevant to their own work made in other institutes.[34] And despite impressive improvements in our information storage and retrieval systems in recent years, the equally impressive increase in the volume of published scientific information inevitably means that there are many accidental failures for potentially interrelated ideas to interact, especially where such ideas emerge in different academic disciplines or are published in different countries or in different languages.

Emergence of new ideas may be inhibited not only by barriers to recombination of old ideas but also by a paucity of old ideas that can be recombined or by an exhaustion of relevant potentialities within a given frame of reference; that is, by a state of affairs in which all relevant combinations have already been conceived or attempted. The ending of progress in the development of oared ships appears to provide an outstanding example. Oared ships coexisted with sailing ships but were used for somewhat different purposes: sailing ships for long-distance commercial voyages and oared ships for sudden speed and maneuverability in combat. The development of oared ships began with Phoenician and Greek boats that were open and had a row of oarsmen (ultimately as many as twenty-five) on each side. Later oared ships became much larger, with many more oarsmen arranged in two or three banks (stories) and with as many as eight men pulling each oar. Eventually such ships came to be powered by the muscles of thousands of oarsmen. But this line of development inevitably exhausted its potentialities, as further increases in numbers of oarsmen made ships increasingly unwieldy and aggravated problems in ship design. Innovations intended to pack still more oarsmen aboard came to an end, and further progress involved a shift toward totally new power sources to replace human muscles and wind-driven sails.[35]

The preceding discussion has suggested the following broad possibilities: technological innovation may be *self-limiting* if it satisfies old needs without creating new ones or if it exhausts the range of possible relevant combinations of items within a given frame of reference, *self-reinforcing* if it creates new needs while satisfying old ones or if it creates new technological items which can be combined with each other and with older items to produce still newer ones. The following additional considerations may also be relevant:

1. New ideas can generally emerge more readily than new practices, not only because the idea for a new practice is logically prerequisite to the institution of the new practice itself, but also because an idea can exist and be used as a basis for generating still newer ideas even if only a very few people know about it and accept it, whereas many practices (e.g., "rules of the road") must be standardized on a society-wide basis if they are to be instituted at all. However, this difference between ideas and practices is more relevant in some contexts than in others. In a small primitive society in which sociocultural and occupational differentiations are minimal and in which people tend to think alike as well as act alike, new ideas may appear just as hard to establish as new practices. At the opposite extreme, there may be situations in which drastic novelties in both ideas and practices are strongly encouraged. The distinction in question is most likely to assume greatest importance when a new idea, despite widespread opposition to it or disinterest in it, can nevertheless find a home in a differentiated segment of society (e.g., in the case of a scientific idea, in a specialized scientific community), whereas many proposed new practices that require broad societal support for their establishment do not receive such support and are thereby blocked.

2. The relative importance of "needs" and "opportunities" as

stimuli for innovation has occasionally been a subject of dispute. The prolonged failure of medical researchers to find a dramatically simple cure for cancer has been attributed by some observers to resistance by vested interests that do not want a cure (i.e., in our current terminology, to a need among some people to preserve the current situation). However, this seems unlikely because the cancer problem is a global one whereas relevant vested interests in the status quo tend to be geographically localized and because the possibility of an immense prestige reward provides a powerful incentive for whoever might find a cure. When numerous brilliant and eager researchers widely distributed around the globe fail for a long time to solve a particular technological problem despite considerable financial support, this failure may best be attributed to the intrinsic difficulty of the problem itself, that is, to limited opportunities for a solution, rather than to any need to keep the problem unsolved. However, in a society that has a generally static technology (as distinct from one that is technologically dynamic but has its progress sharply limited in only a few areas), a powerful need for stability is likely to be more important than any paucity of potentially available opportunities as an inhibitor of technological innovation. In fact, we can recognize retrospectively that a number of important scientific discoveries that required no special instruments could have been made through simple observation and logic long before they actually were made—they "could have been" made, that is, on the basis of opportunities available if strong needs for stability had not intervened.

3. Strong resistance to technological change within particular societies and groups is often overcome by external forces including competitive pressures. Thus, as Lewis Mumford noted, wars stimulate innovations that armies resist.[36] Resistance to technological changes appears relatively important when we examine social situations *micro*scopically (e.g., the internal structures of armies over a short period of time) but appears much less important when we shift to a *macro*scopic view (e.g., the history of warfare over the centuries).

4. Business corporations and government bureaus that sponsor research aimed at producing technological innovations have a direct interest in finding ways to maximize the efficiency of such research. Accordingly, there has emerged a large body of specialized literature on conditions influencing innovation. "Research on research" in this sense has appeared not only in the West[37] but also in the Soviet Union, where it aims specifically at the goal of permitting research productivity to continue its exponential increase even after the resources devoted to research can no longer increase exponentially.[38] Studies in this

area, as one might expect, focus on innovations of a sort that businesses and governments are most likely to be interested in controlling and able to control, and on controllable environmental conditions. The study of sources of innovation within this framework is reasonably advanced with a solid core of knowledge going clearly beyond what common sense would suggest. However, when we broaden our focus of attention to include *all* technological innovations in the sense in which this term has been used here, all potentially influencing conditions, all cultural settings, the picture changes sharply. Our understanding of technological innovation in such a broadened frame of reference is very limited, despite scattered historical and anthropological studies and despite a concentration of inquiry in the related but nevertheless distinct area of the "psychology of creativity."

7
Impacts of Technology

The Study of Impacts

In the broadest sense, "impact assessment" encompasses the investigatin of consequences, effects, or "impacts" of any phenomenon on society or on its environment. Several distinctions among forms of impact assessment in this sense may be noted:

1. A distinction is commonly made between "environmental" and "social" impacts. This distinction will be ignored here for two reasons: (a) the basic methodological issues related to impact assessment arise equally in connection with both forms and (b) the distinction between the two forms is not a clear one: important environmental impacts entail social impacts as well.

2. Impact assessment may be historical or retrospective, concerned with the past (e.g., "What was the initial impact of the first use of the steam engine?") or may be predictive or anticipatory ("What will be the impact of precise and accurate earthquake prediction?") or may be concerned with contemporary situations that extend both backward and forward in time ("What has been, is, and will be the impact of computers?"). Problems inherent in any attempt to predict the future make anticipatory impact assessments particularly difficult in one respect. In another respect, however, historical impact assessments may be more difficult. Impacts of the past tend to become taken for granted and thus tend to lose their status as generally recognized impacts. In an age of hydrogen bombs and space travel, we cannot easily appreciate the nineteenth-century view of the *railroad* as "an

enormous, an incalculable force . . . let loose suddenly on mankind"[1], although this was indeed a reasonable assessment. Alexis de Tocqueville offered a cogent explanation for the systematic forgetting of major past impacts: "Great successful revolutions, by effecting the disappearance of the causes which brought them about, by their very success, become themselves incomprehensible."[2] Future impacts may be difficult to predict but, because they will change situations that currently exist, may often be more readily comprehensible than historical impacts that have changed earlier and no-longer-remembered situations.

In addition to the preceding distinctions, we may also note the following specific forms of impact assessment:

1. In "functional analysis" the investigator seeks to identify ways in which a social phenomenon contributes to the survival of the society (or other social system) in which it occurs, and thus contributes to its own survival. For example, a functional analysis of the Hopi rain dance suggests that this dance, which is regarded by the Hopi themselves as a means for producing rain, actually performs a very different function instead, helping to sustain societal morale and unity and contributing in this way to the survival of Hopi society and hence to its own survival.[3]

Functional analysis focuses largely on certain particular kinds of impacting phenomena and certain particular kinds of impacts. The impacting phenomena may be aspects of particular societies (e.g., the Hopi rain dance, the American political machine[4]) or may be universal or nearly universal aspects of societies in general (e.g., the division of labor, social inequality[5]: the incest taboo); but in any case they tend to be stable and long enduring and are often not technological as that term is used in this book. A functional analysis focuses specifically on those impacts that provide a partial basis for explaining the persistence of the impacting phenomenon itself, that is, on those impacts that have positive rather than negative implications for the survival of the system of which the impacting phenomenon is a part.[6]

2. Cost-effectiveness analysis and cost-benefit analysis both divide the impacts or effects of the phenomenon being analyzed into two component parts: "costs," and "benefits." In cost-effectiveness analysis both of these are quantified but not necessarily in the same terms; we could say, for example, that a certain program is expected to save

one life for every X dollars spent, without deciding how much a life is worth in dollars. In cost-benefit analysis, however, costs and benefits are compared to each other directly, with costs subtracted from benefits to yield an overall evaluation (e.g., subtracting the cost in dollars from the dollar value of lives saved). This requires that costs and benefits be measured by the same criterion, usually a monetary one.[7]

Cost-benefit analysis is sometimes regarded as distinct from "impact assessment" rather than as a form of the latter. Thus, one investigator writes that cost-benefit analysis "is designed for problems where the impact is broad but shallow" whereas impact assessment is designed to cope with "the narrow and deep impact."[8] We are working here with a somewhat different conception according to which impact assessment is a general phenomenon encompassing cost-benefit analysis as a special form.

Cost-benefit analysis is superficially similar to functional analysis in that both involve distinctions between positive and negative impacts. However, the differences between the two kinds of analysis are profound. In functional analysis, impacts are classified as positive or negative according to whether they help or hinder the survival prospects of the impacting phenomenon; the investigator is concerned only with those impacts that are positive in this sense; and the purpose is to explain the impacting phenomenon. In cost-benefit analysis, by contrast, impacts are classified as positive or negative according to whatever values the investigator introduces (on his/her own behalf or on behalf of a client or employer); positive and negative impacts are taken into account together; and the purpose of the analysis is to evaluate the impacting phenomenon rather than to explain it.

3. "Technology assessment" is a form of impact assessment distinguished simply by the fact that the impacting phenomenon is technological. Sometimes technology assessment utilizes cost-benefit concepts, but this is not an essential characteristic. There may also be some overlap between technology assessment and functional analysis, but *typical* cases of each of these stand in sharp contrast to each other. Functional analysis focuses mostly on enduring and self-reinforcing impacts of stable nontechnological phenomena, whereas technology assessment focuses largely on disruptive implications of recent and anticipated technological innovations.

The sources or conditions of technological innovation, and the impact of technology, may be equally important *theoretically* but there is a great difference between them with respect to *practical* importance. The conditions under which technological innovation is relatively likely to occur have broad relevance for countries that are competing with other countries economically and/or militarily in technological achievement; but identification of these conditions does not appear to have direct and acute significance except for planners and administrators of research programs in universities, business corporations, governmental agencies, and private foundations. Furthermore, the practical importance of distinguishing between those conditions that tend to generate technological innovations and those which do not, is reduced by two considerations: (1) Given a long enough time and a wide enough variety of human situations, any technological form that is consistent with the realities of nature and with basic features of human nature is likely to occur sometime and somewhere. (2) Once a technological innovation has made its appearance, it may then be diffused to other settings—even to those in which conditions for original innovation have been highly unfavorable. The *impact* of new technology, by contrast, is quite reasonably a matter of fundamental and universal interest: new technology may create massive unemployment and/or open up major new employment opportunities; it may dramatically raise living standards—and may pave the way for the extinction of the human species via warfare with modern weapons or via damage to the natural environment.

The investigation of technological impacts differs from the investigation of sources of technological innovation not only in its greater practical importance but also in that it is legally mandated in a variety of contexts. This has been true in the United States especially since passage of the National Environmental Policy Act of 1969 (Public Law 91–190), which generally prohibited Federal agencies from engaging in activities with important environmental effects without the prior filing of "environmental impact statements." Inquiries into sources and conditions of technological innovation generally lack comparable legal significance. Even when not legally mandated, assessment of impacts has become a familiar and a routine process in many government agencies and business corporations, much more so than analysis of sources or conditions of technological innovation.

The Conquest of Smallpox: A Case Study

Certain aspects of the historical development of human control over smallpox illustrate complexities that commonly arise in technology assessment.

Smallpox is an ancient disease: the remains of Egyptian Pharoah Ramses V (died ca. 1160 B.C.) suggest that he suffered from it. However, clarification of distinctions between smallpox and other ailments was not easily achieved. Around A.D. 900 the Persian physician Phazes distinguished smallpox from measles; but almost eight centuries later this distinction was apparently not clear to one author, Thomas Thatcher, who published in 1677 "A Brief Rule to Guide the Common-People of New-England how to order themselves and theirs in the Small Pocks, or Measels."[9] The distinction between smallpox and chicken pox was not clarified until 1767.

Highly contagious and deadly, smallpox was especially devastating when introduced into populations that lacked immunity to it; e.g., Iceland in 1707, where smallpox killed eighteen thousand people in a total population of fifty-seven thousand.[10] However, two features of smallpox made it potentially among the easiest infectious diseases to control: survivors acquired immunity; and the disease was limited to humans only, with no possibility of reinfection from animal sources after all human cases were brought under control.

Long before the age of modern medicine began, parents would sometimes deliberately give their children mild cases of smallpox so that they would thereafter be protected against the disease in its severer forms. However, this was a dangerous process in several ways: sometimes other infections would be transmitted accidentally along with smallpox; sometimes a child who was being deliberately given a mild case of smallpox would actually contract a severe case instead; and a person who was protected from smallpox after recovering from a (mild or severe) case of it would nevertheless be a source of contagion for others during the time of his illness. Artificial protection against smallpox was so dangerous that many parents were undecided about whether to attempt such protection for their children. One father who hesitated was Benjamin Franklin, with results that he discussed as follows:

> In 1736, I lost one of my sons, a fine boy of four years old, by the smallpox taken in the common way. I long regretted bitterly, and still regret, that I had not given it to him by inoculation. This I mention for the sake of parents who omit that operation on the supposition that they should never forgive themselves if a child died under it—my example showing that the regret may be the same either way . . .[11]

Near the dawn of the nineteenth century, a safer method for protecting people against smallpox emerged. The English physician

William Jenner, inspired by the comment of a milkmaid that she was safe from smallpox because she had already recovered from "cowpox," discovered a way of conferring immunity to smallpox by giving people cowpox, a much less dangerous disease, through a process that came to be known as "vaccinnation" (although one recent investigator has claimed that Jenner's "cowpox" was really a mild variant of smallpox itself).[12]

In 1800 Dr. Benjamin Waterhouse wrote about "A Prospect For Exterminating Smallpox," a prospect that was first becoming realistic around that time. As the nineteenth century progressed, millions of people in technologically advanced countries—but still only a small fraction of the total world population—came to be protected by vaccination.

Not everyone was pleased by advances against smallpox. Long before there was much progress against it, religious objections were raised. Thus, in 1721 a critic of medical experimentation in Boston insisted that smallpox is "a judgment of God on the sins of the people" and that "to avert it is . . . an encroachment on the prerogatives of Jehovah, whose right it is to wound and smite."[13] Exactly a century and a half later, in 1871, the great biologist Charles Darwin (1809–82) expressed concerns of a very different sort.

> There is reason to believe that vaccination has preserved thousands who from a weak constitution would formerly have succombed to smallpox. Thus the weak members of civilized society propagate their kind. No one who has attended to the breeding of domestic animals will doubt that this must be highly injurious to the race of man[14]

Arguments such as this, popular in the nineteenth century, have become generally less convincing in the twentieth. Observers today would be more likely to refute Darwin by pointing out that conquest of smallpox entails a change in the meaning of "weakness"; if a person was "weak" in relation to smallpox, being highly susceptible to that disease, this condition ceases to be a weakness once smallpox has been eradicated.

The struggle to exterminate smallpox proceeded relatively slowly until 1967, when the World Health Organization launched a coordinated international campaign against that disease. Although a total of 181 years was required after Jenner's development of an effective vaccination system in 1796 before the last naturally occuring case of the disease was found, nevertheless when the end for smallpox actually came, it came with stunning swiftness. In 1974 India reported

218,000 cases; the *last* case in India was reported May 24, 1975. Bangladesh in April 1975 had smallpox cases in 1208 villages; the last case in Bangladesh was found on October 16 of that same year.[15] In October 1976 an author writing in *Scientific American* suggested that "the world may have seen its last case of the most devastating disease in human history." Actually, the last *natural* case was found in October 1977 in Africa. But then, in February 1979, the *New York Times Magazine* ran a story under the heading, "Smallpox Is Not Dead."

What had happened was this. The baccillus that produces smallpox had become extinct *in nature*. However, small collections continue to be kept alive artificially in carefully guarded laboratories in the United States, Britain, and the USSR. The status of the smallpox baccillus today is thus roughly comparable to the status of an animal species that is extinct in the wild but continues to thrive in a few zoos.

Authorities have been reluctant to destroy the remaining captive baccilli for several reasons: such action would constitute a presumably irreversible destruction of a living species (not a generally approved type of action these days); it might preclude some useful research projects in the future; and—more ominously—smallpox is "the ideal biological warfare agent since it is stable, easily aerosolized, simple to grow, and is a terrifying disease with high lethality."[16]

In September 1978, several months after smallpox had become extinct in nature, a medical photographer in England died of smallpox after being accidentally exposed to deliberately cultivated smallpox bacilli. Future accidental exposures or use of smallpox in biological warfare or the evolution of a similar organism (e.g., some other kind of animal pox) into a new kind of smallpox in response to the "vacuum" created in nature by the disappearance of natural smallpox would very likely find huge populations no longer protected either by previous exposure to the disease in its natural form or by vaccination, which, in the absence of the disease, has ceased to be widely practiced. The worst smallpox epidemics in the past have come after the disease has been largely absent for a number of years (e.g., in Boston in 1721, after that city had been smallpox-free since 1702), and thus after levels of immunity have fallen off. We might now be in the midst of a clear period to be followed ultimately by a new and ultradeadly epidemic for which we are becoming increasingly unprepared both in terms of immunities and in other ways.

Basic Problems In Technology Assessment

Attempts to identify the impacts of particular technological phenomena are likely to encounter problems of the following sorts:

1. In addition to direct primary impacts, which are relatively easy to identify, there are likely to be indirect secondary and tertiary impacts that may be nonobvious and harder to uncover. For example: (a) Dry rice cultivation among the Tanala of Madagascar exhausted the soil and thus required that people move to new locations every few years, whereas wet rice cultivation involved permanent use of the same land and hence a settled life with permanent villages. The shift from dry rice to wet rice cultivation accordingly had drastic effects on diverse institutions far removed from agriculture (e.g., warfare, because fixed villages could construct defensive fortifications that periodically moving communities could not construct).[17] (b) Food-labelling legislation in Hawaii enabled people there to learn that fishcakes they had routinely eaten contained shark meat, which led to a decline in sales of these fishcakes, a reduced demand for shark meat, a reduced effort by fishermen to catch sharks, and an apparent increase in the shark population in Hawaiian waters[18]. (c) Alvin Toffler states that in the early twentieth century, increased use of the automobile in the United States produced an increased demand for Brazilian rubber, which was satisfied partly by the enslavement of Brazilian Indians, of whom about thirty thousand reportedly died between 1900 and 1911.[19] (d) Lynn White has identified major features of medieval European society as indirect consequences of the adoption of the stirrup.[20]

2. Especially in times of rapid change, impacts of one innovation may be hard to disentangle from impacts of others. How shall we distinguish, for example, among the effects of the automobile, the telephone, and the airplane, all of which emerged very roughly around the same time?

3. Impacts of different innovations and events may not only reinforce each other and be hard to disentangle, as suggested above, but may also counteract and nullify each other. Eyeglasses, invented in about the 1280s A.D., made it possible for men whose work required detailed vision to work more years than had been possible previously and thus tended to delay promotion of younger men; but this consequence was delayed for many decades by social turmoil and by the Black Death of 1348–49, which apparently opened up new promotion opportunities for youth and thus "masked the reverse effects of the

introduction of eyeglasses."[21] In recent years there has been scientific speculation about the possibility that air pollution may have opposing effects on climate, with carbon dioxide released into the atmosphere tending to create warmer weather while particulate pollution makes the weather colder, with each of these tendencies masking the other, and with the ultimate outcome remaining in doubt.

4. The impact of a given technological innovation may vary enormously in different social settings. For example, printing in China helped to produce greater standardization and uniformity in written works, whereas printing in Europe stimulated an explosive development of new and "heretical" ideas. Perhaps, when differences between societies are profound and enduring, as the differences between China and Europe have been, we could develop concepts that would enable us to understand and predict the differing implications of new technology for each of these societies. However, we must allow for two additional kinds of complexities:

a) People and groups experiencing certain changes may acquire "immunities" to what would otherwise be impacts of similar changes in the future. A major contrast between nineteenth and twentieth century scientific communities illustrates this possibility. In the nineteenth century, the assumption was commonly made that established scientific theories—notably Newtonian mechanics—constituted a permanently firm foundation for science. Thus, Albert A. Michaelson, whose research helped to destroy the Newtonian system that had dominated physics for more than two centuries, actually insisted, before this result had been achieved, that nothing like it would ever happen: "The more important physical laws and facts of physical science have all been discovered, and these are now so firmly established that the possibility of their ever being supplanted in consequence of new discoveries is exceedingly remote."[22] The overthrow of the Newtonian system in the early twentieth century came as a tremendous shock to those who had expected it to endure forever. But now that scientists generally understand that no theoretical system can reasonably be considered as permanently established, a similar upheaval could no longer have the same kind of impact. There are also grounds for believing that people in modern societies have become increasingly adapted to the prospect of continual change and thus increasingly immune to some of the impacts of changes in earlier times. One of Galileo's prominent contemporaries rejected his discoveries of new "planets" because the seven days of the week had been named for the seven planets (which also corresponded to the seven metals and the seven openings in the head, etc.) and because "if we increase the number of planets, this whole system falls to the ground."[23] Today a new planet would hardly cause many people in modern societies to feel that any system to which they were committed was thereby being undermined.

b) In some cases comparatively minor "chance" variations in circumstances, which are essentially unpredictable, may drastically affect the impact of a technological innovation on society. Depending on such variations, nuclear weapons may lead either to permanent peace through a "balance of terror" or to worldwide destruction of civilization in a nuclear holocaust. Similarly, our control over smallpox may lead, as suggested above, either to permanent liberation of humanity from the suffering that that disease has caused over thousands of years of human history or to a recurrence of smallpox that our loss of immunity could cause to be exceptionally devastating.

5. New technologies which ultimately become much more potent than older competing technologies are often *not* more potent *initially*. Thus musketeers did not, at first, have a clear advantage over archers. As one observer noted in 1590, archers (unlike musketeers) were not loaded down with heavy equipment, did not have to worry about their weapons overheating or their ammunition getting wet, and could shoot several arrows before a musketeer could fire a single bullet. Also, horses hurt with arrows "fall a-yerking, flinging and leaping" . . . [and thus] they do disorder one another"[24] while horses shot with bullets would collapse less dramatically and without disordering others. Similarly, the first railroad trains could not be counted on to win races with horses; commercial jet planes in their early years had problems which sometimes made them less attractive to passengers than older propeller planes; the atomic bombs dropped on Hiroshima and Nagasaki were less destructive of human life than the conventional bombing raid on Dresden, Germany earlier in World War II; and, in the sixteenth century, the concept of a moving Earth was not clearly superior to the concept of the Earth as a stationary center of the universe.

6. In addition to the various empirical problems in predicting technological impacts as noted here, there is also a fundamental *conceptual* problem. Identification of the impact of a technological item implies a comparison between a situation in which this item is present and a situation in which it is absent. However, there may be more than one situation of the latter sort, which could be reasonably employed for comparative purposes; and the outcome of our inquiry may depend heavily on an essentially arbitrary choice among the different possible bases of comparison. Consider, for example, the problem of assessing the impact of television viewing on children. If

children did not view television, they would do something else instead; and the impact on them of television viewing would thus depend on the nature of the alternative activities in which they might otherwise have engaged. However, there is no fully satisfactory procedure for the systematic identification of such alternatives and for deciding which shall be employed for comparative purposes.

Problems in Cost-Benefit Analysis

Cost-benefit analysis entails, on top of all problems inherent in impact assessment generally, a variety of additional problems associated with the classification of impacts as "beneficial" or as "costly" and the attempt to specify in standardized quantitative terms the precise amounts of benefits and costs that are involved.

The values with which cost-benefit analysis is concerned are often intangible. Indirect methods for the quantitative measurement of intangible values have been developed. For example, the value that society assigns in practice to a human life can be measured by comparing the wages of occupations that are similar in every important respect except that they involve sharply differing and easily measurable degrees of risk to the lives of their practitioners.

However, there remain difficult technical problems with such attempts to measure values that are commonly considered unmeasurable. The comparisons that these attempts call for may not always be feasible; for example, we may not be able to find two occupations in the same society or community that are really similar in all important respects except for a sharp and measurable difference in the degrees of risk that they entail. And allowance must be made for complex variations in values among different segments of society and among the same people under differing conditions.

Certain additional difficulties with the quantitative assessment of intangible values in cost-benefit analysis are more serious because they involve questions of principle rather than merely "technical" considerations:

1. Attempts to measure intangible values quantitatively may tend to *lower* such values (for example, we might be cheapening human life if we assigned a dollar value to it).

2. The quantitative assessment of costs and benefits tends to exclude consideration of *rights* and *duties:* as Steven Kelman notes, "We would not permit rape even if it could be demonstrated that the rapist derived enormous happiness from his act while the victim

experienced only minor displeasure"[25] (i.e., even if a cost-benefit analysis that took into account only the total amounts of pleasure and displeasure involved should show a positive balance on the side of pleasure).

As this latter example suggests, quantitative assessments of costs and benefits are complicated by conflicting interests. To consider another, more complex, example: minimum wage laws may be beneficial to employed workers and their families, costly to employers, and costly to people who remain unemployed because no one can afford to hire them at required wage levels; there is no simple way to classify high wages as a "benefit" or a "cost" to society as a whole. Beneficial results "for the greatest number" may appear to provide a helpful guideline in some situations; but this criterion does not give us a satisfactory basis for making decisions (1) when the welfare of unborn future generations depends on sacrifices by the generations now living, (2) when decision-making authorities have certain legally or morally binding commitments to give top priority to the needs of their own constituents (fellow citizens in the case of a national leader, shareholders in the case of a corporation executive) rather than to the welfare of humanity as a whole, or (3) when action that is best for most people is only mildly beneficial to them but has exceedingly severe negative implications for a minority.

The cost-benefit distinction is complicated further in some historical contexts by a paradoxical and often-overlooked tendency for almost everything that happens to *become* increasingly beneficial and less costly with the passage of time. I became acutely aware of this tendency when, after I had summarized the immensely destructive consequences of World War II in one of my classes, a student said: "My parents met while my father was serving in the army during the war. If the war had never occurred, my parents would never have met, and I would never have been born—so the war was certainly good for me." In fact, probably anyone born after the war could say that "if the war had never been fought, I would not have been born"—which means that the war was a good thing from the standpoint of all born after it, and it will eventually become a good thing for everyone on earth. Of course, if there had been no war, other people would have been born instead; and all of them would have had good reason to be thankful that the outbreak of war was averted; but it is hard for us to adopt the perspective of purely hypothetical people who were never even conceived because their "parents" never met.

Assumptions about what is primarily beneficial and what is primarily costly are often made without careful attention to the complex issues involved. A particularly interesting example of such an assumption concerns the *reversibility* of technological arrangements. Problems associated with *ir*reversibility are well known. Thus, in one public transportation system, buses were ordered with the specification that windows were to be sealed so that they could not be opened. This was done to prevent children from poking their arms out the windows and to prevent passengers from interfering with the air-conditioning system. But this also meant that when the air-conditioning did not work passengers had no other way of obtaining ventilation; and they suffered, especially in hot weather.[26] And a more important example may be cited: as the First World War was about to begin, the Kaiser sought to reverse the deployment of the German Army from the western front (against France) to the eastern front (against Russia) and was told (incorrectly, as it later turned out) that the deployment in a westerly direction "cannot be altered."[27] Cases such as these have led several investigators to stress the virtues of reversibility. One author has suggested that "highly irreversible decisions should be avoided insofar as possible."[28] This point of view was also explicitly adopted in a National Academy of Sciences report to the House of Representatives Committee on Science and Astronautics:

> a basic principle of decision-making should be to maintain the greatest practicable latitude for future action. Other things being equal, those technological projects or developments should be favored that leave maximum room for maneuver. The reversibility of an action should thus be counted as a major benefit: its irreversibility a major cost.[29]

The advantages that follow as a consequence of reversibility may seem intuitively obvious. Yet we should also consider the implications of the fact that in various spheres of life other than material technology, reversibility is sometimes feared rather than desired. In the American political system, certain basic ideas (in particular, those formulated in the Bill of Rights) are incorporated directly into the United States Constitution rather than merely in ordinary legislation, precisely because we citizens do *not* want them to be easily reversible. In the democracy of ancient Athens, which did not have a difficult-to-amend constitution in which cherished principles could be fairly safely embedded, the irreversibility of certain legislation was sought by providing severe penalties for anyone who proposed that it be reversed: thus, the death penalty was prescribed in 431 B.C. for anyone suggesting that a certain special fund be used for any purpose other than to defend the city against an attack from the sea.[30] Our legal system makes it relatively easy for people to commit themselves to specified courses of action "irreversibly" through the signing of contracts; a contractual agreement that is too easily reversible may thereby lose much of its value. The movement away from irreversibility in marriage is widely regarded as a social problem. Why, then, should irreversibility, which is sought in so many other contexts, be considered a defect when it appears in material technology? There may be a good reason, but the burden of proof falls on those who insist that reversibility in technology is a valid general principle, and they have hardly proved their case. What I am criticizing here is not the conclusion that some investigators have reached favoring reversibility in material technology, but rather the *way* this conclusion has apparently been reached, without an attempt to explain why material technology should be expected to differ from various other phenomena in this respect. And this failure, in turn, reflects an unwarranted barrier that separates the study of material technology from studies of other technologies and of other nontechnological aspects of human society.

8
Conclusion

Technology has been defined here to encompass tools and practices deliberately employed as natural (rather than supernatural) means for attaining clearly identifiable ends. This definition *ex*cludes much seemingly "technological" behavior among lower-animal species *and* human practices that are embedded in unquestioned tradition and thus do not constitute "means" employed to attain identifiable "ends" *and* practices aimed at attaining ends through supernatural means. The definition is broad enough, however, to encompass various organizational arrangements and symbolic systems as well as material tools and machines. It means that some practices that involve rejections of advanced material technology, or that are employed when such technology is not available, nevertheless retain a "technological" character. For example, the reliance on highly coordinated masses of laborers instead of on advanced machinery in many work situations in Mao's China involved an emphasis on "organizational" rather than on "material" technology, not a rejection or absence of technology itself. Our definition also means that some technologies may remain in existence while ceasing to *be* "technologies" and that semitechnological phenomena will often be encountered. The United States Constitution and the alphabet have been discussed here to illustrate these possibilities.

This conception of technology has been linked in the present study with a particular conception of society and its transformations: a conception that emphasizes a distinction between societal "development" and societal "evolution," that interprets evolutionary concepts as more fundamental than developmental concepts, that identifies technological innovation as a decisively important source of social-evolutionary change, and that makes major use of technological criteria in differentiating among social-evolutionary stages.

98

Primitive technologies were generally simple, based on use of human muscle, created through trial and error rather than through systematic inquiry, excellently adapted to local environments, and compatible with small populations and with minimal division of labor. The long historical transition from primitiveness to modernity did *not* involve movement with respect to all these features simultaneously and at the same speed. For example, in several of the great civilizations of the ancient world, which were intermediate between primitiveness and modernity, the primitive reliance on human muscles remained largely unchanged even though small primitive work groups had been superseded by giant organizations in which the muscles of thousands of workers were coordinated. Progress toward increased modernity appears, paradoxically, to entail contrasting trends at different levels: new medical treatments are often simpler than old ones; scientists seek simplicity (parsimony) in their theories; new automobiles are simpler to drive than older ones were—and yet the overall technological system that encompasses medical treatment, scientific theorizing, and automobile driving has been moving not toward greater simplicity but in an exactly opposite direction.

Analyses of technological modernization may be complicated also by two additional considerations. First, the circumstances of societies that completed the shift to modernity some time ago and that pioneered the modernization process are drastically different from the circumstances of "transitional" societies today, which are modernizing via rapid importation of technologies from abroad. Second, technology in each country (or each cultural region) is organized partly in accordance with local styles, and the resulting stylistic variations should be distinguished from indicators of differences in levels or stages of modernization. An examination of technology in China today in Chapter 5 has illustrated both the nature of transitional societies and the concept of "national style" in technological organization.

Recent years have seen greatly increased concern both with the conditions under which technological innovation is most likely (and least likely) to occur and with the impact of technology on society. In both of these areas of inquiry, we encounter a variety of research problems that have been discussed in the preceding pages, and also two general weaknesses in our knowledge, which may be noted here:

1. Different perspectives and research traditions in these areas are only very imperfectly integrated. Scholarly studies and action-oriented studies are largely separated by conceptual barriers that impede potentially fruitful interaction between them. Studies of technology and societal evolution are similarly separated from studies of technology

and socioeconomic development. Studies of the conditions under which societies and civilizations produce creative technological innovations are largely separated from studies of the conditions of technological creativity within research laboratories in which people are hired to be innovative. Impact studies undertaken within the conceptual framework of functional analysis are generally treated as if totally unrelated to impact studies within the frameworks of cost-benefit analysis and technology assessment. Contemporary-impact studies are often conducted without adequate utilization of insights that have emerged from more historically oriented impact studies. Material technology is often treated as if unrelated to nonmaterial and especially to organizational technology.

2. Studies of social conditions that make technological innovation possible and studies of social impacts of technology have yielded considerable information pertaining to *particular* types of situations. For example, the manager of a research organization who wishes to increase productivity will find a considerable and rather impressive body of literature that deals precisely with that problem. However, if we seek *general* knowledge about the conditions under which technological innovation is likely to occur, covering all forms of this phenomenon in all countries and covering variables beyond a manager's control (e.g. "national character") as well as those which may be manipulable (personnel policies), the situation is quite different: available knowledge of that sort is meager indeed. Similarly there is an impressive body of literature on the impacts of certain particular technologies, for example, large-scale irrigation, whereas there remain many equally important technologies that have apparently never had their impacts carefully analyzed. But this state of affairs in the study of technology in society is characteristic of social science as a whole, which has achieved impressive successes in numerous small areas scattered widely across an immense landscape remaining largely unexplored.

Notes

I. INTRODUCTION

1. J. B. S. Haldane, *Possible Worlds* (London: Chatto and Windus, 1930), p. 296.

2. Clifford C. Furnas, *The Next Hundred Years: The Unfinished Business of Science* (New York: Reynal & Hitchcock, 1936), p. 266.

3. *Popular Mechanics,* July 1909, vol. 12 #1, p. 15

4. Simon Newcomb, "The Outlook For the Flying Machine," *The Independent,* vol. 55, Part 2, October 22, 1903, cited in Noel de Nevers, ed. *Technology and Society* (Reading, Massachusetts: Addison-Wesley, 1972), p. 152.

5. Roger Bacon, *Epistola de secretis operibus,* chapter 4, cited in A. C. Crombie, *Medieval and Early Modern Science,* vol. 1 (Garden City, New York: Doubleday, 1959), p. 55.

6. Statement by I. I. Prezent, cited by David Joravsky, *The Lysenko Affair* (Cambridge, Massachusetts: Harvard University Press, 1970), p. 80.

7. Bernard Barber, *Science and the Social Order* (New York: Free Press, 1952), pp. 109–10.

8. N. Inozemtsov, *Contemporary Capitalism: New Developments and Contradictions* (Moscow: Progress Publishers, 1974), p. 61.

9. Karl von Clausewitz, *On War,* ed. Anatol Rapoport (Harmondsworth, Middlesex, England: Penguin Books, 1968), pp. 102–3.

10. Cited by Bernard Brodie and Fawn M. Brodie, *From Crossbow to H-Bomb* (Bloomington, Indiana: Indiana University Press, 1963), p. 70.

11. Benjamin Franklin, *Autobiography,* ed. Carl Van Doren (New York: Viking Press, 1945), p. 598.

12. James Harvey Robinson, *Introduction to the History of Western Europe* (Boston and London: Ginn & Co., 1902), p. 684.

13. Jacques Ellul, *The Technological Society,* trans. Jon Wilkinson (New York: Alfred A. Knopf, 1964).

14. Leslie A. White, *The Science of Culture* (New York: Farrar, Strauss & Cudahy, 1949), p. 365.

15. Cited by Albert H. Teich, *Technology and Man's Future* (New York: St. Martin's Press, 1972), p. 14.

16. Eleanor Atkinson, *The Story of Chicago and National Development* (Chicago: Little, Chronicle Co., 1903, 1909) p. 114.

2. TECHNOLOGY AND ITS FORMS

1. See Victor C. Ferkiss, *Technological Man: The Myth and The Reality* (New York: George Brazilier, 1969), p. 27.

2. John Kenneth Galbraith, *The New Industrial State* (Boston: Houghton-Miffllin, 1967), p. 12.

3. Maurice N. Richter, Jr., *The Autonomy of Science: A Historical and Comparative Analysis* (Cambridge, Massachusetts: Schenkman, 1981), p. 15.

4. Ibid.

5. Jane Goodall, "Tool-Using and Aimed Throwing in a Community of Free-Living Chimpanzees," *Nature* 201, #4926 (March 28, 1964): 1264–66.

6. Lynn White, Jr., *Medieval Religion and Technology: Collected Essays* (Berkeley: University of California Press, 1978), p. 1.

7. Jacques Ellul, *The Technological Society,* trans. John Wilkinson (New York: Alfred A. Knopf, 1964), p. xxv. Ellul's term is *la technique,* translated into English as *technique,* but this is presumably intended as equivalent to *technology.*

8. David M. Freeman, *Technology and Society: Issues in Assessment, Conflict and Choice* (Chicago: Rand McNally, 1974), p. 5.

9. A situation similar to this, involving the Hopi raindance, is discussed by Robert K. Merton, *Social Theory and Social Structure* (New York: Free Press, 1957), pp. 64–5.

10. A distinction between tools and machines has been emphasized by Lewis Mumford, in *Technics and Civilization* (New York: Harcourt, Brace and World, 1963), in *Technics and Human Development* (New York: Harcourt, Brace and World, 1967), and in *The Pentagon of Power* (New York: Harcourt, Brace and World, 1970). Here, however, we may simply regard machines as tools of a certain kind and thus ignore this distinction.

11. Mumford, *Technics and Civilization.*

12. William H. McNeill, *The Shape of European History* (New York: Oxford University Press, 1974), p. 159.

13. National Academy of Sciences, *Technology: Processes of Assessment and Choice,* A Report to the Committee on Science and Astronautics, House of Representatives (Washington: Government Printing Office, July 1969), p. 16.

14. See Han Suyin, "The Sparrow Shall Fall," *New Yorker,* 10 October 1959, pp. 43–50.

15. Cited in Carl Van Doren, *Benjamin Franklin* (New York: Viking Press, 1938), p. 209.

16. This does not mean that the Supreme Court, if called upon today, would rule that a state could adopt a European-style governmental system. Given the large body of precedents that the Court would have to respect, it would very likely find a reason to rule against such a system. However, a fresh examination of the Constitution that ignores the precedents accumulated through nearly two centuries of its operation does not reveal any clear and direct incompatibility between the original Constitutional document and a European-style republican government at the state level.

17. For a discussion of relevant features of the United States Constitution, see Samuel P. Huntington, "Political Modernization: America vs. Europe," *World Politics* 18, no. 3 (April 1966): 391–408.

18. Cited by George F. Foster, *Se-quo-yah: The American Cadmus and Modern Moses* (Philadelphia: Office of the Indian Rights Association, 1885; republished by AMS, 1979), p. 92.

19. For more detailed descriptions of Sequoyah's achievement, see Foster, *op. cit.;* Grant Foreman, *Sequoyah* (Norman: University of Oklahoma Press, 1938); and Dale Van Every, *Disinherited* (New York: William Morrow, 1966), pp. 62–72.

3. SOCIETY AND ITS EVOLUTION

1. As of May 1982, Syrian troops and their local allies control part of Lebanon's capital, Beirut, and surrounding areas; Maronite Christians control other parts of Beirut and of the Lebanese countryside; the Palestine Liberation Organization controls a large part of southern Lebanon; and an Israeli-supported Lebanese general controls a narrow strip of Lebanon along the Israeli border.

2. Arnold J. Toynbee, "History," in R. W. Livingstone, ed., *The Legacy of Greece* (Oxford: at the Clarendon Press, 1921), p. 290, cited by Robert A. Nisbet, *Social Change and History: Aspects of the Western Theory of Development* (London: Oxford University Press, 1969), p. 217.

3. Oswald Spengler, *Decline Of the West,* 2 vols. (New York: Alfred A. Knopf, 1939).

4. For excellent discussions of the concept of evolution and its applicability to social situations, see Donald T. Campbell, "Variation and Selective Retention in Socio-cultural Evolution," in Herbert R. Barringer, George I. Blanksten, and Raymond W. Mack, eds., *Social Change in Developing Areas* (Cambridge, Massachusetts: Schenkman Pub. Co., 1965), pp. 19–49, and Donald T. Campbell, "Evolutionary Epistemology," in P. A. Schilpp, ed., *The Philosophy of Karl Popper* (Lasalle, Illinois: Open Court, 1974), pp. 413–63. For discussions of the historical background of theories of progress or social evolution, see John B. Bury, *The Idea of Progress: An Inquiry Into Its Origin and Growth* (London: Macmillan 1928); Frederick J. Teggart, *The Idea of*

Progress (Berkeley: University of California Press, 1929); Frederick J. Teggart, *Theory and Processes of History* (Berkeley: University of California Press, 1941); Robert A. Nisbet, *op. cit.*, and Kenneth Bock, "Theories of Progress, Development, Evolution," in Tom Bottomore and Robert Nisbet, eds., *A History of Sociological Analysis* (New York: Basic Books, 1978).

5. Cited by Nisbet, *op. cit.*, p. 66.

6. Lewis H. Morgan, *Ancient Society* (New York: Henry Holt, 1877), new edition edited by Leslie A. White (Cambridge, Mass.: The Belknap Press of Harvard University Press, 1964).

7. Friedrich Engels, *The Origin of the Family, Private Property and the State* (New York: International Publishers, 1942).

8. Lewis Morgan, *op. cit.*, p. 522.

9. The relation between Marxist and Christian conceptions of progress is discussed by Nisbet, *op. cit.*

10. Ferdinand Tonnies, *Community and Society*, trans. Charles P. Loomis (East Lansing, Michigan: Michigan State University Press, 1957).

11. Henry Maine, *Ancient Law* (London: Oxford University Press, 1946).

12. Howard Becker, *Through Values to Social Interpretation* (Durham, North Carolina: Duke University Press, 1950), ch. 5, "Sacred and Secular Societies," pp. 248–280.

13. Robert Redfield, *The Primitive World and Its Transformations* (Ithaca, New York: Cornell University Press, 1953).

14. Emile Durkheim, *The Division of Labor in Society*, trans. George Simpson (New York: Free Press, 1966).

15. Herbert Spencer, *On Social Evolution*, ed. J. D. Y. Peel (Chicago: University of Chicago Press, 1972).

16. Talcott Parsons, *The Evolution of Societies*, ed. Jackson Toby (Englewood Cliffs, New Jersey: Prentice-Hall, 1977). To Parsons "differentiation" is only one of four "main types of structural change" involved in societal evolution: the others, which need not be discussed here, are called "adaptive upgrading," "inclusion," and "value generalization." However, evolutionary types of society are distinguished by Parsons on the basis of "differentiation" alone.

17. Leonard T. Hobhouse, *Morals In Evolution* (London: Chapman & Hall, 1951), "Types of Society," pp. 38–69.

18. Lewis Mumford, *Technics and Civilization* (New York: Harcourt Brace and World, 1963).

19. Gerhard Lenski and Jean Lenski, *Human Societies: An Introduction to Macrosociology*, 4th ed. (New York: McGraw-Hill, 1982). See also Gerhard Lenski's earlier but more detailed work, *Power and Privilege* (New York: McGraw-Hill, 1966).

20. Quantitative comparisons of pay scales and incomes in capitalist and socialist societies do not provide an adequate basis for drawing conclusions about relative amounts of inequality in the two kinds of society. Among other considerations that one must take into account are special privileges that high-ranking people may have, which defy quantification. In the Soviet Union, for

example, such privileges may include permission to travel in the West, access to uncensored or minimally censored news from the outside world, use of special traffic lanes. In addition, the centralization of decision making in the Soviet Union entails extremely great inequalities in the distribution of power, and this also does not show up in comparative studies of the distribution of quantitatively measurable rewards such as income or pay.

21. This point has been discussed in Richard Rubinson, "The World Economy and the Distribution of Income Within States: A Cross-National Study," *American Sociological Review* 41:638–659; Maurice N. Richter, Jr., "Stratification At a Global Level," paper presented to the American Sociological Association, (New York: August 1976); and Maurice N. Richter, Jr., *Society: A Macroscopic View* (Boston: G. K. Hall, and Cambridge, Mass.: Schenkman, 1980), p. 13.

22. Adam Ferguson, *An Essay On the History of Civil Society* (Edinburgh: Edinburgh University Press, 1966, first published 1767), p. 169.

23. For additional discussion of issues mentioned here, see W. F. Wertheim, *Evolution and Revolution: The Rising Waves of Emancipation* (Harmondsworth, Middlesex, England: Penguin Books, 1974); also Maurice N. Richter, Jr., *Society: A Macroscopic View* pp. 34–36.

24. See Immanuel Wallerstein, *The Modern World System*, 2 vols. (New York: Academic Press, 1974, 1980); Andre Gunder Frank, "The Development of Underdevelopment," *Monthly Review* 18 (September, 1966): 17–31; Rudolfo Stavenhagen, *Social Classes in Agrarian Societies* (Garden City, New York: Doubleday Anchor, 1975).

25. Parsons, *op. cit.*, p. 215.

26. For a detailed discussion of these events, see J. H. Parry, *The Age of Reconnaissance* (New York: Mentor, 1964).

27. This historical process is nicely described by William H. McNeill, *The Rise of the West: A History of the Human Community* (Chicago: University of Chicago Press, 1963).

28. Cited by A. C. Parker, *New York State Museum Bulletin* 613 (1913), and by Anthony F. C. Wallace, "Culture and Cognition," *Science* 135 (February 2, 1962), p. 356.

29. The rise of modern science has been described and analyzed by many investigators. I have summarized some essential features of this development in *Science As a Cultural Process* (Cambridge, Massachusetts: Schenkman, 1972), and in *The Autonomy Of Science: A Historical And Comparative Analysis* (Cambridge, Massachusetts: Schenkman, 1981).

30. Lewis H. Morgan, *op. cit.*, p. 468.

4. DIRECTIONS OF TECHNOLOGICAL CHANGE

1. See Asen Balikci, *The Netsilik Eskimo* (Garden City, New York: Natural History Press, 1970), and J. Garth Taylor, *Netsilik Eskimo Material Culture* (Oslo: Universitetsforlaget, 1974).

2. Repeated movements of a hunting-and-gathering society in response to local depletion of resources is illustrated by the Mbuti pygmies of the Congo, who find that "after about a month . . . the fruits of the forest have been gathered all around the vicinity of the camp, and the game has been scared away. . . ." (Colin Turnbull, "The Mbuti Pygmies of the Congo," in James Gibbs, ed., *Peoples of Africa* [New York: Henry Holt, 1965], pp. 286–7.) A similar pattern has prevailed in some much more advanced agrarian societies, but involving only a comparatively small number of societal members whose luxurious lifestyles have been incompatible with geographical stability; thus, "We see sovereigns and great nobles all through the Middle Ages, traveling from one estate to another with [their] ministers and trains: eating up the year's produce in a week or few days, and then passing on to eat up a fresh estate." (George G. Coulton, *Medieval Panorama* [New York: Macmillan, 1938], p. 233).

3. Balikci, *op. cit.*

4. Karl A. Wittfogel, *Oriental Despotism: A Study of Total Power* (New Haven: Yale University Press, 1957).

5. Vernard Foley and Werner Soedel, "Ancient Oared Warships," *Scientific American* 244, no. 4 (April 1981): 148–63.

6. Bernard Brodie and Fawn M. Brodie, *From Crossbow to H-Bomb*, revised (Bloomington, Indiana: Indiana University Press, 1973).

7. Foley and Soedel, *op. cit.*

8. Margaret Mead and Frances Cooke MacGregor, *Growth and Culture: A Photographic Study of Balineses Childhood* (New York: G. P. Putnam's Sons, 1951), p. 42.

9. See Goodall, *op. cit.*, pp. 1264–66.

10. Washington to Arthur Young, December 5, 1791, cited by Daniel J. Boorstin, *The Americans: The Colonial Experience* (New York: Random House, 1958), p. 260.

11. Lewis Thomas, *The Lives of a Cell: Notes of a Biology Watcher* (New York: Viking Press, 1974), p. 32.

12. William Crookes, in *Fortnightly Review,* 51:173–81, 1892, cited by James R. Bright, "Technological Forecasting Literature: Emergence and Impact on Technological Innovation," in Patrick Kelly and Melvin Kranzberg, *Technological Innovation: A Critical Review of Current Knowledge* (San Francisco: San Francisco Press, 1978), 299–334, p. 309.

13. Siegfried Giedion, *Mechanization Takes Command* (New York: Oxford University Press, 1948), p. 50.

14. Maurice N. Richter, Jr., *The Autonomy Of Science: A Historical and Comparative Analysis* (Cambridge, Massachusetts: Schenkman, 1981), pp. 138–9.

15. Wolfram Fischer, "The Role of Science and Technology in the Economic Development of Modern Germany," in William Beranek, Jr. and Gustav Ranis, eds., *Science, Technology and Economic Development: A Historical and Comparative Study* (New York: Praeger, 1978), p. 75.

16. This information on the Franklin stove comes from Gerald Holton, in *Daedalus* 109, no. 1 (Winter 1980): 10.

17. For a discussion of self-correcting social-organizational and political arrangements, see Allen Schick, "The Cybernetic State," *Transaction* 7, no. 4 (February 1970): 14–26.

18. William Foote Whyte, "Social Inventions for Solving Human Problems," *American Sociological Review* 47, no. 1 (February 1982) p. 5.

19. S. Colum Gilfillan, *Supplement to 'The Sociology of Invention'* (San Francisco: San Francisco Press, 1971), p. 40.

20. This tendency was discussed by William Fielding Ogburn, *Social Change With Respect to Culture And Original Nature* (New York: Viking Press, 1930), pp. 104–107.

21. Lewis Mumford, *The Pentagon Of Power,* (New York: Harcourt, Brace and World, 1970).

22. Friedrich von Bernhardi, *How Germany Makes War* (New York: George H. Doran Co. 1914) p. 95.

23. Statements of the Chief of Cavalry, March 11, 1940, Hearings Before the Subcommittee of the Committee on Appropriations, House of Representatives, 67th Congress, 3rd Session, on the Military Establishment Appropriation Bill for 1941, pp. 678–9.

24. See S. Colum Gilfillan, *Inventing the Ship* (Chicago, Follett, 1935).

25. E. F. Schumacher, *Small Is Beautiful: Economics as if People Mattered* (New York: Harper & Row, 1973), p. 34.

26. Division of Intergovernmental Science and Public Technology, Directorate for Engineering and Applied Science, National Science Foundation, *Appropriate Technology: Summary of Awards, Fiscal Year 1980* (Washington: National Science Foundation, 1981).

27. Harvey Brooks, "A Critique of the Concept of Appropriate Technology," in Franklin A. Long and Alexandra Oleson, *Appropriate Technology and Social Values: A Critical Appraisal* (Cambridge, Massachusetts: Ballinger, 1980) 53–78, p. 65.

28. Thorstein Veblen, *The Engineers and the Price System* (New York: Viking Press, 1954) p. 57.

29. Recuperative capacities of modern industrial societies have been commented on by Kenneth E. Boulding, *The Meaning of the Twentieth Century* (New York: Harper & Row, 1964).

5. VARIATIONS IN TECHNOLOGICAL STYLE: THE CASE OF CHINA

1. I am most grateful to the American Enterprise Institute for Public Policy Research for providing the financial support that made this trip possible. The trip was taken while I was in residence at AEI with a National Endowment for the Humanities fellowship.

2. There is no contradiction here: two countries may have inherited quite

different cultural traditions from the distant past, while at the same time they share important cultural features that have emerged in both countries as a consequence of similar contemporary problems.

3. Marco Polo, *The Book of Sir Marco Polo* (New York: Airmont Pub. Co., 1969), first written A.D. 1298.

4. For a longer list see Charles Singer, " 'Epilogue' East and West in Retrospect," in Charles Singer *et. al.,* eds., *A History of Technology* (New York: Oxford University Press, 1956) 2:770–71. For a detailed description of ancient Chinese technology, see Joseph Needham, *Science And Civilization in China* (Cambridge: Cambridge University Press, 1951–).

5. Cited by Arnold J. Toynbee, *A Study of History,* abridgment by D. C. Somervell (New York: Dell Pub. Co., 1965), 1:55.

6. See Ralph C. Crozier, *Traditional Medicine in Modern China* (Cambridge, Massachusetts: Harvard University Press, 1968), and E. Grey Dimond, *More Than Herbs and Acupuncture* (New York: W. W. Norton, 1975).

7. See National Academy of Sciences, *Acupuncture Anaesthesia in the People's Republic of China, A Trip Report of the American Acupuncture Anaesthesia Study Group* (Washington: Committee For Scholarly Communication With the People's Republic of China, 1976).

8. Rudi Volti, "The Absorption and Assimilation of Acquired Technology," in Richard Baum, ed., *China's Four Modernizations: The New Technological Revolution* (Boulder, Colorado: Westview Press, 1980), 179–201, p. 195.

9. From Mao Zedong, *On New Democracy* (Beijing: Foreign Languages Press, 1960).

10. See Haicheng Earthquake Study Delegation, "Prediction of the Haicheng Earthquake," *EΘS (Transactions of the American Geophysical Union* 58, no. 5 (1977): 236–72; Frank Press *et. al.,* "Earthquake Research in China," *EΘS (Transactions of the American Geophysical Union)* 56, no. 11 (1975): 838–81; and Edward C. T. Chao, "Earth Sciences," in Leo Orleans, ed., *Science in Contemporary China* (Stanford, California: Stanford University Press, 1980), 190–212.

11. Edward C. T. Chao, *op. cit.*

12. Victor K. McElheny, "Total Synthesis of Insulin in Red China," *Science* 153, no. 3733 (July 15, 1966): 281–3.

13. Dun J. Li, *The Ageless Chinese* (New York: Charles Scribner's Sons, 1978) p. 124.

14. Bohdan O. Szuprowicz, "Electronics," in Leo Orleans, *op. cit.,* 435–61, p. 444.

6. TECHNOLOGICAL INNOVATION

1. For a brief account of this episode, see Derek J. de Solla Price, *Science Since Babylon* (New Haven: Yale University Press, 1961).

2. Thucydides, *History of the Peloponnesian War,* trans. Rex Warner (Baltimore: Penguin Books, 1972), p. 313.

3. Bernard Brodie and Fawn M. Brodie, *From Crossbow to H-Bomb* (Bloomington: Indiana University Press, 1973), p. 24.

4. See ch. 2, notes 18 and 19.

5. See Joseph Ben-David, *The Scientists Role in Society* (Englewood Cliffs, New Jersey: Prentice-Hall, 1971), pp. 139–142.

6. Boris M. Hessen, "The Social and Economic Roots of Newton's Principia," in George Basalla, ed., *The Rise of Modern Science* (Lexington, Massachusetts: D. C. Heath, 1968), pp. 31–38.

7. Ralph Linton, *The Study of Man* (New York: Appleton-Century, 1936).

8. National Science Foundation, *Federal Funds for Research, Development and Other Scientific Activities: Fiscal Years 1976, 1977, and 1978* (Washington: National Science Foundation, 1978—NSF 78-300 vol. 26 [GPO], p. 48.

9. National Science Foundation, *National Patterns of R & D Resources: Funds and Personnel in the United States, 1953–1978/79* (Washington: National Science Foundation, 1978—NSF78-813, GPO 1978), p. 25.

10. Donald C. Pelz and Frank M. Andrews, *Scientists in Organizations: Productive Climates For Research and Development* (Ann Arbor, Michigan: Institute for Social Research, University of Michigan, 1976).

11. Ben-David, *op. cit.*

12. William H. McNeill, *The Rise of the West: A History of the Human Community* (Chicago: University of Chicago Press, 1963), p. 273.

13. Thus, if the Soviet Union rather than the United States had become the world center of scientific activity in the post–World War II era, this outcome could have been explained in a very "reasonable" way: the Soviet emphasis on cooperation rather than on competition, on group rather than on individual activities, is more compatible with contemporary trends in large-scale science than is the individually-competitive emphasis prevailing in the United States. Because postwar Soviet achievements, however impressive they may be in view of the initial Soviet handicap, nevertheless fail to come close to matching American achievements in science, this argument is not made, except in propagandist Soviet or Soviet-oriented writings. Instead, a totally different argument is made to explain the actual American scientific preeminence.

14. For good examples from the sociology of science, see Bernard Barber, *Science And the Social Order* (New York: Free Press, 1952), and Robert K. Merton, *The Sociology of Science* (Chicago: University of Chicago Press, 1973). Science appears to need "freedom" most badly when its fate in antiscientific dictatorships (e.g., Nazi Germany) is under examination. However, when attention focuses instead on dictatorships that impose heavy restrictions on freedom but that are fundamentally proscientific in various other respects, the status of freedom as a prerequisite for science may sometimes appear to be a more complicated matter. For a more detailed discussion of this question, see Maurice N. Richter, Jr., *The Autonomy of*

Science: A Historical and Comparative Analysis (Cambridge, Massachusetts: Schenkman, 1981).

15. Joseph Ben-David, *op. cit.*

16. Northrup Frye, "The Bridge of Language," *Science* 212, no. 4491 (April 10, 1981): 127–36, p. 128.

17. Maurice N. Richter, Jr., *op. cit.*, p. 167.

18. Derek J. de Solla Price, *op. cit.*; also Price, *Little Science, Big Science,* (New York: Columbia University Press, 1963.)

19. Cited by Nicholas Rescher, *Scientific Progress* (Oxford: Basil Blackwell, 1978), p. 54.

20. Cited by Rescher, *op. cit.*

21. Friedrich Engels, *Dialectics of Nature,* 3rd ed. rev. (Moscow: Progress Publishers, 1964), pp. 22–23.

22. Stacy V. Jones, *Inventions Necessity Is Not The Mother Of* (New York: Quadrangle Books, 1971).

23. Cited by H. G. Barnett, *Innovation: The Basis of Cultural Change* (New York: McGraw-Hill, 1953), p. 249.

24. Cited by C. A. H. Smith, "The Discovery of Anesthesia," *Scientific Monthly* 24 (January 1927): 64–70.

25. Cited from *The Grand Concerns of England* (1673) by Dietrich Schroder, *Physics And Its Fifth Dimension: Society* (Reading, Massachusetts: Addison-Wesley, 1972), p. 21.

26. This criticism of Galileo by Francesco Sizi is cited by Gerald Holton and H. D. Roller, *Foundations of Modern Physical Science* (Reading, Massachusetts: Addison-Wesley, 1958), p. 160.

27. An 1897 statement cited by John B. Rae, "The Internal Combustion Engine on Wheels," in Melvin Kranzberg and Carroll W. Pursell, Jr., *Technology In Western Civilization* (New York, Oxford University Press, 1967), 1, 119–137, p. 123.

28. Bernard Berelson and Gary A. Steiner, *Human Behavior: An Inventory of Scientific Findings* (New York: Harcourt, Brace and World, 1964), p. 614.

29. Sidney Cohen, *The Drug Dilemma* (New York: McGraw-Hill, 1969), p. 71.

30. Henri Poincaré, *The Foundations of Science,* trans. George Bruce Halsted (Lancaster, Pennsylvania: The Science Press, 1946), p. 387.

31. Derek Price, *Science Since Babylon,* (New Haven: Yale University Press, 1961).

32. See Stephen F. Mason, *A History of the Sciences* (New York: Collier, 1962).

33. For a few years in the early 1950s, there was an unusually strong emphasis in the United States on secrecy in militarily relevant areas of science. Events related to this phenomenon are discussed by Edward Shils in *The Torment of Secrecy* (New York: Free Press, 1956). Generally, secrecy in science has been much more profound in the Soviet Union, where all scientific research is considered secret unless and until permission for open publication is granted.

34. See Maurice N. Richter, Jr., *The Autonomy of Science*, Chapter 7, "The Chinese Science System," 131–156.

35. See Vernard Foley and Werner Soedel, "Ancient Oared Warships," *Scientific American* 244, no. 4 (April 1981): 148–63.

36. Lewis Mumford, *Technics and Civilization* (New York: Harcourt, Brace, and World, 1963), p. 95.

37. For a good sampling of studies in this area see Jacob Schmookler, *Innovation and Economic Growth* (Cambridge, Massachusetts: Harvard University Press, 1966); Neil M. Kay, *The Innovating Firm: A Behavioral Theory of Corporate Research and Development* (New York: St. Martins, 1976); Donald M. Pelz and Frank M. Andrews, *op. cit.*; Patrick Kelly and Melvin Kranzberg, eds., *Technological Innovation: A Critical Review of Current Knowledge*, (San Francisco: San Francisco Press, 1978); and Frank M. Andrews, *Scientific Productivity: The Effectiveness of Research Groups in Six Countries* (Cambridge: Cambridge University Press, 1979).

38. See I. Y. Sheinin, *Science Policy: Problems and Prospects*, (Moscow: Progress Publishers, 1978); also D. M. Gvishiani, S. R. Mikulinsky, and S. A. Kugel, *The Scientific Intelligentsia in the USSR* (Moscow: Progress Publishers, 1976).

7. IMPACTS OF TECHNOLOGY

1. Charles Adams, Jr., 1868, cited in Richard C. Dorf, *Technology, Society and Man* (San Francisco: Boyd and Fraser, 1974).

2. Alexis de Tocqueville, *L'Ancien Régime*, trans. M. W. Patterson (Oxford: Basil Blackwell, 1952) p. 7.

3. Discussed by Robert K. Merton, *Social Theory and Social Structure* (New York: Free Press, 1957), pp. 64–5.

4. Discussed by Merton, *op. cit.* p. 71–82.

5. The functional analysis of social inequality and critiques of such analysis are summarized and reviewed by George A. Huaco, "The Functionalist Theory of Stratification: Two Decades of Controversy," *Inquiry* (Oslo, Norway) 9 (1966): 215–240.

6. There are several mechanisms by which the "functions" of a social phenomenon may facilitate the survival of that phenomenon itself. For an examination of issues involved in this aspect of functional analysis, see Maurice N. Richter, Jr., "Social Functions and Sociological Explanation," *Sociology and Social Research* 50, no. 1 (1965).

7. This distinction between cost-effectiveness analysis and cost-benefit analysis is discussed by Peter H. Rossi and Sonia Rosenbaum, "Program Evaluation," in Marvin E. Olsen and Michael Micklin, eds., *Handbook of Applied Sociology* (New York: Praeger, 1981), 91–114, p. 104.

8. Peter G. Sassene, "Social Impact Assessment and Cost-Benefit Analysis," in Kurt Finsterbusch and C. P. Wolf, eds., *Methodology For Social Impact Assessment* (Stroudsberg, Pennsylvania: Dowden, Hutchinson, and Ross, 1977), 74–82, p. 79.

9. Cited by Ola Elizabeth Winslow, *A Destroying Angel: The Conquest of Smallpox in Colonial Boston* (Boston: Houghton Mifflin, 1974), p. 26.

10. Donald A. Henderson, "The Eradication of Smallpox," *Scientific American*, 235, no. 4 (October 1976): p. 27.

11. Benjamin Franklin, *The Autobiography and Other Writings*, ed. L. Jesse Lemisch (New York: New American Library, 1961), p. 112.

12. Peter Razzell, *Edward Jenner's Cowpox Vaccine: The History of a Medical Myth* (Firle, Sussex, England: Caliban Books, 1977).

13. Cited by Andrew D. White, *A History of the Warfare of Science With Theology in Christendom* (New York: George Brazillier, 1955, first published 1895), p. 56.

14. Charles Darwin, *The Descent of Man* (New York: D. Appleton, 1897) p. 134.

15. Henderson, *op cit.*, p. 32.

16. Nicholas Wade, "Biological Warfare Fears May Impede Last Goal of Smallpox Eradicators," *Science* 201 (July 28, 1978): p. 330.

17. Ralph Linton, *The Tanala: A Hill Tribe of Madagascar* (Chicago: Field Museum of Natural History, 1957).

18. Edmund S. Hobson and E. H. Chave, *Hawaiian Reef Animals* (Honolulu: University Press of Hawaii, 1972), p. 9.

19. Alvin Toffler, *The Third Wave* (New York: Bantam Books, 1980), p. 90.

20. Lynn White, *Medieval Technology and Social Change* (Oxford: Clarendon Press, 1946), p. 28.

21. Lynn White, Jr., "Technology Assessment From the Standpoint of a Medieval Historian," *American Historical Review* 79, no. 1 (February 1974): 1–13, pp. 5–6.

22. A. A. Michaelson, *Light Waves and their Uses* (Chicago: University of Chicago Press, 1961), pp. 23–24.

23. Francesco Sizi, cited by Gerald Holton and H. D. Roller, *Foundations of Modern Physical Science* (Reading, Massachusetts: Addison-Wesley, 1958), p. 160.

24. Sir John Smythe, *Certain Discourses Military*, ed. J. R. Hale, (Ithaca, New York: Cornell University Press 1964, first published 1590) p. 76.

25. Steven Kelman, "Cost-Benefit Analysis: An Ethical Critique," *Regulation* 5, no. 1 (January/February 1981): p. 35. See also the replies to Kelman in *Regulation* 5, no. 3 (May/June 1981): pp. 2–3, and Kelman's response to his critics.

26. Albert R. Karr, "Bus Windows Sealed Shut Boil Riders," *Wall Street Journal,* August 20, 1980, section 2, p. 21.

27. Barbara Tuchman, *The Guns of August* (New York: Dell, 1967), p. 98.

28. David N. Jackson, *Technology and Society: Issues in Assessment, Conflict and Choice* (Chicago: Rand McNally, 1974), p. 133.

29. National Academy of Sciences, "Report of the National Academy of Sciences to the Committee on Science and Astronautics, United States House

of Representatives" (Washington: Government Printing Office, July 1969), p. 32.

30. Thucydides, *History of the Peloponnesian War,* trans. Rex Warner (Baltimore: Penguin Books, 1972), p. 139. Twenty years later, according to Thucydides (p. 546) the Athenians voted to use this fund for another urgent military purpose—*after* voting to cancel the death penalty for suggesting this.

Index

6170